NOFA

Organic Principles and Practices Handbook Series
A Project of the Northeast Organic Farming Association

Organic Seed Production and Saving

The Wisdom of Plant Heritage

Revised and Updated

BRYAN CONNOLLY

C. R. LAWN, Contributing Editor

Illustrated by Jocelyn Langer

CHELSEA GREEN PUBLISHING
WHITE RIVER JUNCTION, VERMONT

Originally published in 2004 as *The Wisdom of
Plant Heritage: Organic Seed Production and Saving.*

"Bee pollinating an echinacea blossom" on page 26
adapted from "Busy Bee" by Wilfrank Paypa on
Wikimedia Commons (http://commons.wikimedia.org/
wiki/File:Frankeys_creation_-_Busy_BEE_%28by%29.jpg).
All other photos by the author.

Editorial Coordinator: Makenna Goodman
Project Manager: Bill Bokermann
Copy Editor: Cannon Labrie
Proofreader: Helen Walden
Indexer: Peggy Holloway
Designer: Peter Holm, Sterling Hill Productions

Printed in the United States of America
First Chelsea Green revised and updated printing March, 2011
10 9 8 7 6 5 4 3 2 1   11 12 13 14

green
press
INITIATIVE

Chelsea Green Publishing is committed to preserving
ancient forests and natural resources. We elected to print
this title on 30-percent postconsumer recycled paper,
processed chlorine-free. As a result, for this printing, we
have saved:

5 Trees (40' tall and 6-8" diameter)
2 Million BTUs of Total Energy
466 Pounds of Greenhouse Gases
2,243 Gallons of Wastewater
136 Pounds of Solid Waste

Chelsea Green Publishing made this paper choice because
we and our printer, Thomson-Shore, Inc., are members
of the Green Press Initiative, a nonprofit program dedicated to supporting authors, publishers, and suppliers
in their efforts to reduce their use of fiber obtained
from endangered forests. For more information, visit:
www.greenpressinitiative.org.

Environmental impact estimates were made using the Environmental Defense Paper Calculator.
For more information visit: www.papercalculator.org.

Our Commitment to Green Publishing

Chelsea Green sees publishing as a tool for cultural change and ecological stewardship. We strive to align our
book manufacturing practices with our editorial mission and to reduce the impact of our business enterprise in
the environment. We print our books and catalogs on chlorine-free recycled paper, using vegetable-based inks
whenever possible. This book may cost slightly more because we use recycled paper, and we hope you'll agree
that it's worth it. Chelsea Green is a member of the Green Press Initiative (www.greenpressinitiative.org), a
nonprofit coalition of publishers, manufacturers, and authors working to protect the world's endangered forests
and conserve natural resources. *Organic Seed Production and Saving* was printed on Joy White, a 30-percent postconsumer recycled paper supplied by Thomson-Shore.

Library of Congress Cataloging-in-Publication Data
Connolly, Bryan.
 Organic seed production and saving : the wisdom of plant heritage / Bryan Connolly ; C.R. Lawn, contributing
editor ; illustrated by Jocelyn Langer. -- Updated.
    p. cm. -- (Organic principles and practices handbook series)
"A Project of the Northeast Organic Farming Association."
"Originally published in 2004 as The Wisdom of Plant Heritage: Organic Seed Production and Saving."
Includes bibliographical references and index.
ISBN 978-1-60358-353-4
1. Seed technology. 2. Organic gardening. 3. Vegetable gardening. 4. Vegetables--Seeds. 5. Vegetables--
Propagation. I. Northeast Organic Farming Association. II. Title. III. Series: Organic principles and practices
handbook series.

SB117.C717 2011
631.5'21--dc22

2011000417

Chelsea Green Publishing Company
Post Office Box 428
White River Junction, VT 05001
(802) 295-6300
www.chelseagreen.com

FSC
www.fsc.org
MIX
Paper from
responsible sources
FSC® C013483

Best Practices for Farmers and Gardeners

The NOFA handbook series is designed to give a comprehensive view of key farming practices from the organic perspective. The content is geared to serious farmers, gardeners, and homesteaders and those looking to make the transition to organic practices.

Many readers may have arrived at their own best methods to suit their situations of place and pocketbook. These handbooks may help practitioners review and reconsider their concepts and practices in light of holistic biological realities, classic works, and recent research.

Organic agriculture has deep roots and a complex paradigm that stands in bold contrast to the industrialized conventional agriculture that is dominant today. It's critical that organic farming get a fair hearing in the public arena—and that farmers have access not only to the real dirt on organic methods and practices but also to the concepts behind them.

About This Series

The Northeast Organic Farming Association (NOFA) is one of the oldest organic agriculture organizations in the country, dedicated to organic food production and a safer, healthier environment. NOFA has independent chapters in Connecticut, Massachusetts, New Hampshire, New Jersey, New York, Rhode Island, and Vermont.

This handbook series began with a gift to NOFA/Mass and continues under the NOFA Interstate Council with support from NOFA/Mass and a generous grant from Sustainable Agriculture Research and Education (SARE). The project has utilized the expertise of NOFA members and other organic farmers and educators in the Northeast as writers and reviewers. Help also came from the Pennsylvania Association for Sustainable Agriculture and from the Maine Organic Farmers and Gardeners Association.

Jocelyn Langer illustrated the series, and Jonathan von Ranson edited it and coordinated the project. The Manuals Project Committee included Bill Duesing, Steve Gilman, Elizabeth Henderson, Julie Rawson, and Jonathan von Ranson. The committee thanks SARE and the wonderful farmers and educators whose willing commitment it represents.

CONTENTS

Acknowledgments

I would like to thank C. R. Lawn for his editing and contributions, and for the assistance, interviews, information, and support I would like to thank Elisheva Rogosa, Rowen White, Harry Records, Michael Glos, Mark Hutton, Tom Stearns, Beth and Nathan Corymb, Jack and Anne Lazor, Diane Dorfer, Jonathan von Ranson, and Bill Duesing. Without their help this manual would not have been possible.

Introduction: Our Heritage

In this time of species extinction, homogenization, corporate gigantism, and the theft of commonly held genetic material as intellectual property—the skills of seed saving are vital for small organic food producers and the whole of agriculture. Farmers and gardeners need to be able to produce their seeds as a basic, inalienable way to control their means of production. And a local organic agroecosystem requires an adapted seed source much as a community requires a memory.

Seed saving was the original skill that brought humanity agriculture. The faithful harvesting, planting, and selecting of wild plants eventually created our domesticated crops, which allowed humanity to transform a wild landscape into a cultivated field. The original landscape supported only 0.1 percent of biomass that humans could consume. After domestication, people were able to transform nearly 90 percent of that biomass into sustenance (Diamond 1997). This interaction of plants and humans, while it displaced wild processes and relationships, created surpluses, transforming societies of nomadic hunter-gatherers into sedentary populations. Once people settled and accumulated material wealth, the surpluses freed some from producing food and allowed a class of artisans, craftsmen, priests, and politicians to develop and technological advancement to occur. As the plants influenced people and civilization, people also kept refining plants. Millennia of cultivation, in different climates and for different uses and taste preferences, yielded a vast array of heirloom and locally adapted varieties among food and fiber crops.

Consequences of Genetic Erosion

Today we are heirs to approximately eleven millennia of human–plant interaction. A flip through the most recent Seed Savers Exchange

Yearbook reveals an astounding diversity. This richness is our heritage, and we should preserve it if only to honor the efforts of seed-saving people throughout the ages. But maintaining this diversity is vital for three main reasons:

- Genetic diversity buffers humanity from starvation, whereas monocultures can lead to catastrophic pathogen outbreaks. Famous examples of such crop failures are the potato famine in nineteenth-century Ireland and the Southern corn blight epidemic of 1969–70 in the United States.
- With our environment changing because of soil salinization, loss of the ozone layer, and global warming, humanity has an interest in maintaining plant species that have the reproductive flexibility to adapt to different conditions.
- Diversity helps to avoid inbreeding depression. Crop varieties that have been maintained as small populations begin to express detrimental recessive genes, and the vigor of the plants decreases.

Sadly, because of corporate consolidation of agriculture and cultural erosion, many varieties have been lost, and many more are in the process of going extinct, which diminishes our cultural heritage and reduces our agricultural options. In 1984, 5,000 non-hybrid varieties of vegetables were available in seed catalogs. By 1998, 88 percent of these were no longer offered (Whealy 1999). Though some of these varieties are still being maintained by amateur seed savers, they are no longer available for most growers. Many have disappeared forever. Of 7,000 apple varieties once available, currently only 1,000 survive (Shiva 2000). Each lost variety is like a book taken from the library and burned. Some say "we have all the genes still," and that may be so, but having all the genes is like having a dictionary: all the pieces are there but a dictionary does not do what *Macbeth* or even a Steven King novel does. Each cultivar is a unique combination crafted by climate, soil, history, and human hands to fill its niche. It is a legacy of human–plant interaction that took thousands of years to shape. A variety is a work of art passed to us from people of the past.

Why Seed Saving by Farmers Is of Critical Importance

Seed saving by farmers and gardeners is fundamental to the preservation and improvement of agricultural biodiversity. The United States Department of Agriculture and several international institutions operate seed banks where large collections of varieties are maintained. Though much needed and vital to genetic conservation, these seed banks have limitations.

Many of the acquisitions are frozen for long-term storage, but freezing halts the evolutionary adaptation of a variety, and it may not develop resistance to new pests and diseases (Bellon et al. 1997). When the seed bank grows a crop out to renew or increase the seed, several varieties of the same species are grown at the same time. Despite the use of isolation techniques, some unwanted crossing can occur. Large collections like these can also incur inbreeding depression and genetic drift because a small number of individuals are grown per variety (Nabhan 1989).

Any given farmer or gardener maintains many fewer varieties at a time than a seed bank. People who grow a few varieties can curate them well

Indigenous subsistence farmers in many regions traditionally use landraces—crop varieties that are heterogeneous (several heights, colors, different disease resistances, etc.) and that produce reliably under variable conditions year to year.

by planting a large number of individuals and paying close attention to the varietal characteristics, thus keeping populations vigorous and retaining the raw material for adaptation. Farmers and gardeners also can select the best plants to improve the crop. Good maintenance will allow a crop to evolve and adapt to new agricultural situations while at the same time keeping its good characteristics true to type.

Seed saving done well is at the heart of regenerating U.S. agriculture. An ideal system would use few pesticides and fossil fuels while delivering the most and best quality food to the people. Our dominant production system, meanwhile, aims for maximizing yield, often at the expense of quality. An item of organic food is generally higher in vitamins, soluble solids, and minerals than its agribusiness equivalent (Long and Reiley 2004). By selecting hardy plants suited to organic farming conditions, we allow more nutritious food to be grown locally instead of burning fossil fuels to ship watery, low-nutrient vegetables across the country.

An heirloom is a crop variety that has been selected for a set of desired characteristics and been handed down from generation to generation.

Heirlooms in Organic Agricultural Systems

Crop plant species are divided into different varietal types: landraces, heirlooms, open-pollinated varieties, and hybrids. A variety is a population of crop plants that is generally uniform for traits or characteristics. Varieties are consistent and can be easily recognized by the characteristics of the plant, fruit, flower, seed, and so on.

A *landrace* is a population of crop plants that is heterogeneous (several heights, colors, different disease resistances, etc.) developed under a specific set of agroecological and socioeconomic conditions. Subsistence farmers in many developing nations and Native Americans traditionally used landraces. Every year the weather brings a new set of farming challenges—whether it's too wet, too dry, too hot, too cold, too buggy, and so on—but a highly variable plant population is bound to have some individuals that yield well. The average harvest may not yield as heavily as with a uniform variety, but the landraces seldom fail and need minimal inputs.

An *heirloom cultivar* (cultivar = cultivated variety) has been handed down from generation to generation (Ashworth 2002). Varieties that originated before 1945 may not be true family heirlooms, but they were developed in and adapted to an earlier agricultural environment similar to that of organic farms, without chemical fertilizers (Nabhan 2002). After World War II, the large surplus of nitrogen left over from bomb making went to produce fertilizer (Nabhan 2002). Heirloom crops and early commercial varieties couldn't

Zucchini varieties.

handle the huge amounts of nitrogen that were available in the new fertilizer; they would lodge (fall over) from too-rapid growth. Breeders therefore bred the current conventional crop plants to take advantage of this highly fertilized environment. Many heirlooms have retained the ability to grow without these intensive fertilizers.

Open-pollinated (OP) varieties are cultivars increased or multiplied through random mating within the population, in which like begets like—though there is always some minor variation. Heirlooms are open-pollinated too, but OP is usually used to describe varieties that are newer than heirlooms.

Hybrid varieties are created by crossing two plant lines. They are not sterile, but their offspring will not breed true. Hybrids are often bred for high production, broad adaptability, uniformity, and to stand up under long-distance shipping—not necessarily for flavor or nutritional value.

one

Strengths and Limitations of Hybrid Varieties

Hybrid seeds are beneficial in many environments. They frequently have good general adaptability, meaning that they perform well in most growing environments. Seed companies find that the seeds sell well to a mass market. They are generally productive and uniform in maturation, a characteristic that's excellent for machine harvest or harvest by large crews. Some have earlier maturity and many incorporate increased disease resistance. Frequently they have been bred to produce tough skins or rinds to reduce splitting, withstand rough handling, and remain unblemished through transport. Because of these traits, hybrids have come into widespread use, displacing many heirloom varieties, particularly among commercial growers.

The relative merit of hybrids and open-pollinated varieties is now a hotly debated topic among sustainable growers. In some crops, such as carrots and broccoli, hybrids would appear to have the clear advantage. There are no good open-pollinated carrot varieties with *Alternaria* resistance and few, if any, open-pollinated broccoli cultivars with the tight beads and domed heads that withstand hot weather and resist fungal diseases in moist conditions. For these reasons, grower Elizabeth Henderson insists that she requires hybrids for her 300-share CSA.

In other crops, the advantages of hybridization are more questionable. Zucchini hybridizers appear to have traded flavor for higher yields, arguably an acceptable trade-off for high-volume commercial growers but not for home gardeners or gourmet marketers. Tomato hybrids frequently produce unblemished fruit, but most lack that old-timey taste that many folks crave. Only one hybrid, 'Sungold', ranks in my top ten varieties for flavor.

Most other crops probably fall somewhere between these extremes. Organic growers often make a place for the best hybrids, but find they want to retain outstanding open-pollinated varieties in their program.

Some hybrids do not perform well in extreme environments where they are not specifically adapted. In some cases they replaced regional heirlooms and open-pollinated varieties that actually performed better in specific areas by their adaptation to local soils, temperature, rainfall levels, pests, and diseases.

Yield is often cited as the reason why hybrids have displaced heirlooms. While many hybrids boast a big productivity advantage, often they also require higher inputs. Not all hybrids greatly outperform OPs. Frank Kutka, a Cornell graduate student, has found some open-pollinated corns that have yields similar to hybrid corns. Even if hybrids are more productive, what are we trading for the higher yields? Many hybrid varieties are not bred for flavor or nutrition, whereas 'Tuscarora' corn from Native Americans in New York State is a low yielder but has superior nutrition (Frank Kutka, pers. comm.).

Uniformity in maturation is often not an advantage for small farmers and gardeners. Successive ripening extends the season over a longer time and may allow steady sale or consumption of the entire crop instead of a stressful, feast-or-famine situation. Heirloom and open-pollinated varieties may be better suited to many small Northeast farms than hybrids.

In one area, open-pollinated varieties have a great advantage over hybrids. Save seed from an open-pollinated variety and it will come true to type in the next generation. Save seed from a hybrid variety and it will not. Instead, the traits of its inbred parental lines will begin to segregate

(Left to right): Brassica napus; F1 hybrid (interspecific); B. rapa ('Tatsoi').

out. It is possible to transform a hybrid variety into a good open-pollinated one by selecting for preferred traits as they segregate out over several generations, but it is a long, arduous process requiring seven to ten years.

For practical purposes, without the willingness to perform such breeding, farmers who use hybrid seed have to keep going back to the seed company each year for fresh seed. Hybrids thus create a dependency on the large multinational seed companies that currently dominate the industry and retain proprietary control over the lines of the hybrids. These lines are considered trade secrets.

Therefore, farmers who wish to save their own seed will prefer to grow open-pollinated varieties. For those with interest in selecting or breeding hybrids, I offer the basic genetics below. Those lacking interest or patience in such subjects may choose to skip the next section.

The Genetics of Hybrids

Hybrid varieties are the product of crossing two lines of plants, producing uniform offspring that have two different sets of genes, one from each parent variety. These plants are genetically heterogeneous within themselves (called *heterozygous*) but among plants they are all the same. A gene is a bit of DNA that controls a trait. An *allele* (*uh-LEEL*) is an alternative

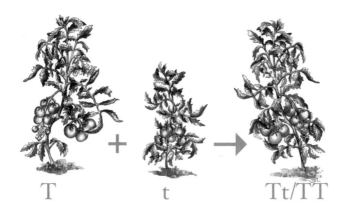

T   t   Tt/TT

Hybrids and characteristics: If two lines of a species are crossed, a dominant characteristic will manifest consistently. In this example of tomatoes it is tallness, but it could as easily be shortness, or fuzziness of stem, etc.

version of a gene. For example, the same gene can have one version that codes for red flowers and another that codes for white. A dominant allele masks the effects of a recessive allele. As an example, let's say red is dominant: if a plant receives the red allele from one parent and a white from the other, the plant will express red. It will express white only if it receives two white alleles, one from each of its parents.

Tomatoes provide a simple example, since they are naturally self-pollinating (also known as inbreeding) and are generally uniform. Considering tall and short genes, if a tomato breeder crosses a short line with a tall line here is what happens:

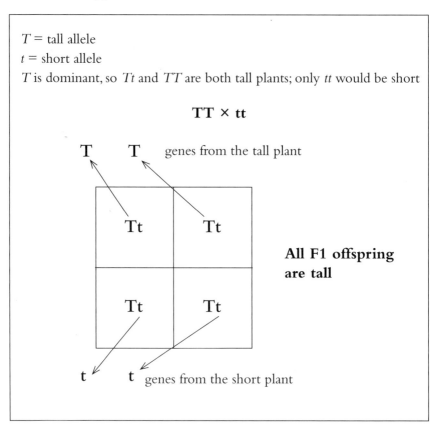

T = tall allele

t = short allele

T is dominant, so *Tt* and *TT* are both tall plants; only *tt* would be short

TT × tt

T   T   genes from the tall plant

Tt   Tt

All F1 offspring are tall

Tt   Tt

t   t   genes from the short plant

All the offspring of this cross are tall—physically or phenotypically uniform—even though within themselves they are heterozygous. Usually when two different lines are crossed to create a hybrid they differ in many genes, some visible and many that we cannot detect. This mixture of genes in the hybrid offspring often gives it "hybrid vigor." The two different sets of genes allow hybrids to deal with many different types of environmental conditions and often just allow them to grow faster. The gene mix is the reason why hybrids can be grown in large areas of the country and are generally more productive.

It is commonly thought that hybrid seeds are sterile. This is generally not true; most hybrid garden varieties do produce viable seeds because they are crosses between two different varieties of the *same* species. Hybrids between *different* species are mostly sterile. For example, *Cucurbita moschata* butternut squash can be crossed with *Cucurbita maxima* hubbard squash fairly easily, but their offspring produce no or very few seeds. The variety 'Tetsukabuto' is a cross between these two species. One very obvious sterile hybrid is seedless watermelon, a cross between two different watermelon varieties that have different chromosome numbers.

Now let's add color to the genetic equation: when a cross is done between plants that differ in two traits, it is called a *dihybrid* cross. Let's say our tall line from above is also a red tomato, and is dominant over yellow, so this line is TTRR, tall and red. Our short line is yellow ttrr. This is what would happen if we crossed them:

$$\text{TTRR} \times \text{ttrr} = \text{TtRr, all tall and red}$$

All the offspring from this cross are uniform (tall and red) among the plants, but, again, have different genes within each plant. If we then mate two TtRr with each other we see F2 segregation, which is the production of offspring with different phenotypes due to the reshuffling of alleles.

From this shuffling of genes in the second generation, called F2 segregation, it can be seen that the commonly held belief of a hybrid reverting to the parental types is only partially true. In the mix of offspring, a reshuffling or segregation of genes has occurred. Because of segregation, the F2 results in parental types and new combinations not seen before. As can be seen in the example we have nine tall red plants and one short yellow plant that are the parental types, but we also have three tall yellows and three short reds—these are new combinations. This F2 segregation is why farmers need to buy new hybrid seed each year. Seeds saved from

Parental alleles				
	TR	**Tr**	**tR**	**tr**
TR	TTRR tall and red	TTRr tall and red	TtRR tall and red	TtRr tall and red
Tr	TTRr tall and red	TTrr tall and yellow★	TtRr tall and red	Ttrr tall and yellow★
tR	TtRR tall and red	TtRr tall and red	ttRR short and red★	ttRr short and red★
tr	TtRr tall and red	Ttrr tall and yellow★	ttRr short and red★	ttrr short and yellow
★new combinations				

(Left margin label: **Parental alleles**)

hybrids will not come true. Hybridization is in effect a copyright or patent. Another company or farmer cannot produce the hybrid unless its original breeder releases its parental lines. A seed-purchasing farmer who finds an excellent hybrid for his growing conditions is therefore always vulnerable. A producer can suddenly drop it, and no one is able to reproduce it. Remember 'Bravo' broccoli, 'Sugar Bowl' muskmelon, and 'Vidi' pepper? R.I.P.

Reclaiming Hybrids

Some innovative farmers and plant breeders such as Alan Kapuler are turning hybrids on their head. They have begun to reclaim varieties from hybrids by stabilizing them and developing open-pollinated varieties from them. They grow seed from an F1 hybrid, allowing segregation to occur in the F2 and subsequent generations. They then select for the desired traits until the seed produces consistent plants with the desired characteristics. Let's take our dihybrid cross, TtRr, from the previous section:

TtRr				
	TR	**Tr**	**tR**	**tr**
TR	**TTRR** **tall and red**	TTRr tall and red	TtRR tall and red	TtRr tall and red
Tr	TTRr tall and red	**TTrr** **tall and yellow★**	TtRr tall and red	Ttrr tall and yellow★
tR	TtRR tall and red	TtRr tall and red	**ttRR** **short and red★**	ttRr short and red★
tr	TtRr tall and red	Ttrr tall and yellow★	ttRr short and red★	**ttrr** **short and yellow**
*★new combinations; **boldface** would breed true if seeds were saved				

If seeds were saved from any of the plants indicated in bold above, their offspring would be stable because they are homozygous, meaning that for each trait they have the same allele (either both capital or both lower case, dominant or recessive), and would now breed true and could form the base for four new open-pollinated varieties. Unfortunately, hybrids often differ in many traits and are unlikely to produce a stable population in two generations, as we see in this example. If seeds were saved from the other plants, their offspring would vary, but over time would eventually stabilize with selection over several generations. Examples of stable varieties derived from hybrids include 'Clear Dawn' onion, which was stabilized from 'Copra', 'Early Moonbeam' watermelon from 'Yellow Doll', and 'True Platinum' sweet corn from 'Platinum Lady'. If you are looking for such a project, tomatoes might be especially easy because they are self-pollinating and need very little isolation distance. Some hybrids may be hybrid in name only and prove to be open-pollinated. Save some seeds and see!

Returning to Resistance

Breeding and selection of open-pollinated varieties has almost been abandoned by academics and seed companies; many universities no longer have vegetable breeders to produce locally adapted new varieties. Some breeders and farmers are again picking up the torch and have begun to use mass selection on open-pollinated varieties. Projects such as the Cornell University and NOFA–NY Public Seed Initiative (PSI) and the Restoring Our Seed (ROS) Project sponsored by USDA SARE (Sustainable Research and Education Program) are currently funded to teach farmers how to select and breed better open-pollinated varieties for their farms.

These farmer-breeder projects are working with a different type of resistance to disease called horizontal resistance, as opposed to vertical resistance. Vertical resistance is the primary resistance type used by commercial and university breeders. Vertical and horizontal have nothing to do with lying down or standing up.

Vertical Resistance

One or a few genes control vertical resistance. The plants are either resistant or susceptible. A source of the resistance must be identified, often by disease-screening thousands of accessions (varieties) or related wild species from a seed bank. In time, diseases tend to overcome or match the vertical resistance and reinfect the crop, though in a few cases vertical resistance has endured for decades. Once it matches the resistance in one plant, this variant of the pathogen will be able to infect all plants with the same genes and break down the resistance. Downy mildew in spinach has repeatedly overcome vertical resistance. Seven matching races of downy mildew have evolved to break down single-gene vertical resistances. Learning from history, mainstream breeders are now breeding for horizontal resistance to this disease. Breakdown has also occurred with bacteria leaf spot in peppers.

Breakdowns happen repeatedly. They are the rule rather than the exception in vertical resistance. Further breakdowns have occurred with several different resistance genes for the same disease in the same species, leading to a variety treadmill, with each new resistance being matched, often within just a few years, especially in tropical regions. For example, in Mexico, varieties of potatoes with vertical resistance to late blight had their genes matched by the pathogen as soon as they sprouted; the pathogen's genetic diversity revealed itself immediately (Robinson 1996).

Horizontal Resistance

By contrast, horizontal resistance relies on multiple genes and multiple mechanisms to prevent infection. Whereas vertical resistance is absolute, horizontal resistance is relative. It can be very slight to almost complete, with all grades in between. Horizontal resistance exists in any cultivar that has a substantial degree of genetic diversity. No specific source of resistance has to be located. If there is enough genetic richness, some degree of horizontal resistance is there. Unfortunately, the widespread use of chemical pesticides may have resulted in erosion of horizontal resistance in the crop-plant world. If there is not high selective pressure for resistant traits, they become lost owing to genetic drift. If the pest is removed, as has been done for years by chemical use, the plant has no reason to maintain its own resistance and becomes susceptible when the threat reappears (e.g., when grown without pesticides or when the pest becomes resistant to pesticides).

The majority of heirlooms and land-races have some horizontal resistance. If these varieties are grown without pesti-cides in an environment with pest and disease pressure, they should have a

Alternaria in brassica.

degree of resilience and be productive under organic farming practices. Further breeding for horizontal resistance is in some ways simpler than breeding for vertical resistance because it employs a straightforward tech-nique—selecting for the healthiest plants. Improvement is gradual from generation to generation. One difficulty is the need for a sufficiently large population. Dr. Raoul Robinson, the author of *Return to Resistance* (1996), suggests 10,000 plants or more. In the Northeast this is not practical for most crops except small grains or legumes such as oats, wheat, or peas.

To breed for horizontal resistance, all plants must be subjected to the pathogen, with the most resistant survivors selected. Otherwise, plants might be healthy only because they escaped contact with the disease. One method to limit the number of escapes and to reduce the amount of field space needed is to grow a large population of seedlings in a nursery, inoculate all plants with disease, and then plant survivors into the field. Selecting in the nursery prevents wasting field space on a large population. The survivors go to seed and their offspring are subjected to the same treatment the following season. This process is repeated until an acceptable level of resistance is reached. In Kenya, where farmers produce two crops of corn per year, breeders achieved an acceptable level of resistance to tropical rust in five to seven years (ten to fifteen generations). This resistance has lasted over thirty years and should continue indefinitely. To bring this technique closer to home, Dr. Mark Hutton in Maine and Jeremy Barker-Plotkin in Massachusetts are both selecting for horizontal resistance to early blight disease in tomatoes.

two

Before You Grow the Seed

Selecting Varieties

Selecting a variety to grow for seed can be a challenge. Heirlooms are often adapted to a specific region, climate, and growing conditions. The best way to select an heirloom is to buy from a seed company in your region. Even if your local company does not produce the seed it sells, it should have evaluated the variety and selected for good performance in your area with your growing season. If you are a member of the Seed Savers Exchange, you can also investigate heirloom or open-pollinated varieties that gardeners are growing in the same general climate as yours. Remember that commercial varieties can disappear, so if you have a winning variety, you may want to save seed even if it appears to be secure in the catalogs. Seed companies are always pushing new varieties and often dropping older ones. Don't get a false sense of security: typically, several companies buy seeds from the same source. If that source halts production, the variety could disappear from commercial production very quickly.

Squash diversity.

In addition to seeds from local northeastern sources, those from the upper Midwest and the Northwest grow well in my garden in Connecticut. I have also successfully grown seeds from Native Seeds/SEARCH in Tucson, Arizona. In their catalog, they indicate high elevation-adapted varieties and desert-adapted varieties. I have been most successful with ones adapted to the cooler higher elevations. Consider all aspects of the agroecosystem when choosing a variety. Will this crop mature in your growing season? Is your area too hot or cold? What diseases and insects affect the crop and how common are they? Is the plant day-length sensitive, i.e., does the crop need short days to flower? (Many tropical varieties of plants cannot *begin* to flower in northern areas until late September or October.)

My wife and I have done extensive experiments in our Connecticut garden. One year we grew about a dozen chili pepper varieties that we had purchased from various local seed companies. Most produced well, but we did not like the eating qualities of some. We also grew several peppers from the Southwest, and the opposite was true, we liked the flavor of many but only a few produced well in our climate. Choosing and seeing for yourself is best. You will want to try a variety in more than one place (replication) on your farm or in your garden because your soil, water, light, and microclimate will vary from spot to spot. Try a variety over a few seasons to test its ability to perform in hot versus cool summers and see how it produces with varying amounts of rain.

If you are growing both for the fresh vegetable market and for seed, you should also trial the variety for acceptance by customers. Some great varieties just don't sell. You may want to see if an heirloom is early enough to compete with hybrids, or if it is productive enough. Does it have good flavor and good appearance? Certain heirlooms may sell in your region. For example, Latino customers will purchase the Calabasa squash. A market may be emerging for Asian, Jamaican, Italian, Polish, or African crops that no one is growing for seed. You might speak with people of these ethnic groups to learn their produce preferences. People like a good

A sample of potato diversity.

story, and tradition sells. Many people who visit Maine or have roots there buy Maine 'Yellow Eye' beans. Rhode Island folks often seek out the Rhode Island 'White Flint' corn.

Older heirlooms inevitably start to branch off and develop into new variants called *strains*. These are not different enough to be called a new variety, but are distinct from other versions. Maybe they were grown in a different region and adapted, or perhaps there was some crossing between different cultivars, or a mutation occurred. So, in addition to testing varieties, if you are particularly interested in one type of crop, you may want to test strains. 'Moon and Stars' watermelon is a good example of a variety with many strains. Most of us in the Northeast grow the round red-fleshed version because it has the shortest time to maturity, though 'Cherokee Moon and Stars', the oblong strain, to me is far superior in taste and flesh texture. Unfortunately, it generally will not mature in the Northeast. You may want to test several versions of the same variety, especially if you are getting seeds from individual seed savers.

An excellent place to start your seed-saving activities is with any heirloom variety from your region. In Connecticut, 'Red Wethersfield' onion is very likely to work. In Maine, 'Yellow Eye' beans and 'Boothby's Blonde' cucumber will be good varieties to grow. Ask your neighbors and friends if they know of anything local that is maintained by a family. Once people know you are interested in heirlooms they will share knowledge and seeds with you. You can also obtain interesting varieties from the United States Department of Agriculture (USDA) Germplasm Resource Information Network (GRIN). This section of the USDA has an online database of seeds collected from all over the world, many of which are still being grown out and maintained. These seeds are free to the public for research or reintroduction purposes. The database can be searched by variety name or by the region the seed was collected from, so you can find varieties adapted to your climate. I have requested varieties from USDA that I could not find commercially.

Saving Seed and Improving Crops

Choosing Your Level of Commitment

There are at least four possible levels of commitment to seed growing. At each step there is an increase in the potential benefits and risks to the grower and the level of responsibility involved.

The simplest level is saving seed for your own use. If something goes wrong you will be the only one affected. You may tolerate a degree of crossing that would not be acceptable were you selling your seed. Your benefits are potential money saved from not having to purchase as much seed, greater independence from seed companies, the ability to preserve a variety for yourself even if it is dropped from commerce, and the good feeling of closing the circle of life from seed to seed.

The next level is to select from your best plants for desired traits to improve your varieties and adapt them to the conditions on your farm. Brett Grohsgal of Even' Star Farm in Maryland has successfully selected arugula and other salad greens for cold hardiness so that he can supply them to his markets year-round. He has developed crack resistance in some of his favorite tomatoes and crossed watermelons for better flavor, appearance, and disease resistance. Such selection has improved the performance and versatility of his crops and enabled him to offer strains that his competitors do not have. They are a hit with his restaurant accounts and CSA members.

You could do the same. Not only would you gain the satisfaction of being mostly independent from the seed companies, but also you could actually trump them by producing better strains that are far more adaptable to your growing conditions. And the gains in adaptability often don't take long. You will see visible improvements in two or three years at the most (Brett Grohsgal, pers. comm.).

If anything gets crossed, or goes wrong, of course, you have more to lose than if you were merely seed saving. In particular, you are at special risk to lose your improved gene lines because there is no one to fall back on—you are the only one who has the material.

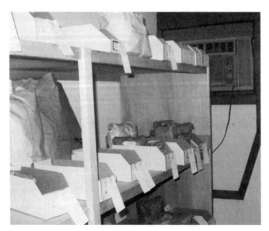

Seed storage room. Note the air conditioner.

Selling Your Seed

The next level is to grow for sale to a seed company. This gives your farm a new marketing niche and a new source of income. It also entails greater risks in committing land and time to the enterprise. You will need a heightened awareness of diseases that must be avoided, and will have to meet certain standards of germination, purity, and seed cleanliness. If anything goes wrong, you affect not only your own income, but the incomes of those growers who are using your seed. However, the seed company to whom you sell the seeds assumes most of the risk.

Contracts

Don't be stuck without a place to sell your seed crop after you have harvested it. Begin by contacting several companies that buy seed (see Resources—Heirloom Seeds and Information). Before you plant, you should have an established agreement with a seed company to ensure an outlet for your product. If you want to grow out an heirloom that a company does not currently sell, you may have to begin two years ahead. The seed company will ask for a seed sample to grow and trial before they decide whether to set up a contract to purchase the seeds.

Economics

When you grow for a seed company you will be selling your seed at a wholesale price (Tom Stearns, pers. comm.). Often you will find that you would make more money retailing your vegetables at farmers' markets. But seed sales may be economically advantageous if your farm or garden is far from a market to sell your crops, or if you lack the time and inclination to hustle markets or deliver your produce. Seeds can be easily sent through the mail. Seed crops also are not as perishable as produce; seed of most crops stored under cool, dry conditions will last several years, allowing enough time to sell the entire crop. If you are selling wholesale, you will receive a payment all at once instead of in several small amounts. Marketing is easier, but the risk of depending upon one or only a few buyers is greater. You could devote a whole summer to a seed crop only to have it fail the germination test.

Seed sales and fresh-market sales may not be mutually exclusive. Some crops can produce a marketable product plus seed, maximizing profits.

Squashes and melons cut in half allow sale of seed plus sale of the fruit wrapped in cellophane. Seeds are removed from broom corn before the tops can be used. There are other possibilities for other crops in certain markets; find what works for you.

What Crops to Sell?

When choosing which crops to sell, consider your climate, your aptitudes, the scale of your operation, the amount of land available, the presence or absence of good isolations (well-separated growing areas to prevent unwanted cross-pollination between varieties), and the market.

All seeds are not created equal. The law of supply and demand prevails. A mainstream variety such as 'Marketmore 76' cucumber is in high demand, and you could probably sell hundreds of pounds of seed. However, you will have to compete with large western growers and can expect to receive only a minimal wholesale price.

The big outfits will not be interested in growing a niche heirloom variety such as 'Boothby's Blonde' cucumber. There is too little demand. If you can land a contract to grow this variety for one of the small seed companies, you can expect to be paid a premium. However, only a few pounds of seed are needed annually to meet the demand, and the market can easily be flooded.

Unique or superior genetics command a higher price on the market. Fedco Seeds pays a premium to farmers who have created open-pollinated varieties out of hybrids or who are growing unique heirlooms otherwise unavailable on the market. Brett Grohsgal is now selling seeds of his superior selected lines at top dollar.

If you have enough available land, you may be able to grow space-grabbers such as *maxima* squash or sweet corn. If yours is a backyard operation, the more compact tomato or pepper varieties may be more appropriate.

Know your seed company markets. If you are growing on a large scale and emphasizing the mainstream varieties, High Mowing Seeds in Vermont or Seeds of Change in New Mexico might be right for you. If yours is a smaller operation geared more to the niche varieties, you might find a good fit growing for Fedco or Turtle Tree. Seeds of Change, High Mowing Seeds, and Turtle Tree sell organic seed only. The latter two particularly appreciate biodynamically grown seed. On the other hand,

Johnny's and Fedco, although they offer a premium price for certified organic seed, also purchase conventionally grown seed crops.

Integrating Seed Crops into a Market Farm

If you are integrating seed crops into a market farm, you will need to develop a plan to keep the competing requirements of these two operations from interfering with each other and allow them, instead, to complement each other to the maximum extent possible. In particular, if you are planning to grow seed for any of the "outcrossing" species (see chapter 3 for a discussion of "selfers" and "outcrossers"), you will need adequate isolations lest your market crops of the same species interfere with your ability to produce a pure seed crop. If you lack adequate land, you might find yourself restricted to growing the selfers for seed. Here are some tricks of the trade that can improve your versatility in this critical area:

(1) Diversification. You can plant one representative of each species without risk of crossing. Thus, among cucurbits, you can plant a *pepo*, a *moschata*, a *maxima*, a watermelon, muskmelon, cucumber, and lagenaria gourd with no crossing.

(2) You can plant seed crops for a multiyear harvest. With proper storage, seeds for a number of species will last two, three, or more years without notable deterioration. Thus you can grow enough 'New England Pie' pumpkin to last two years. This allows you to maintain seed for two different *pepos* instead of one.

(3) Isolations. With even just one isolation, you can double your capabilities for one species. With one isolation and a two-year seed harvest, you will be able to maintain seed for four different *pepos* by alternating the seed crops between the two locations during the two years. Better yet, you can rotate other families into your seed plots if you possess multiple isolations.

(4) Many crops can be harvested both for seeds and for market produce. In his greens plots, Brett Grohsgal

harvests the weaker plants for salad and leaves the stronger as maternal plants for seed production. His "thinnings" become a money-making crop, and he gets two profitable harvests out of the same field.

(5) To improve pollination in his seed-growing areas, Frank Morton arranges particularly compatible plants that bloom together and create a large amount of bee forage but won't cross into seed guilds. Using these and other techniques you can construct an ideal rotation suitable for both market farming and seed production.

Direct-Marketing Your Seed Crops

You might turn your summer avocation into a full-time year-round business. Starting your own seed company or direct-marketing your seeds in any fashion, whether through a garden center, farmers' market, over the Internet, or by a mail order catalog, entails the most risk but also offers the greatest potential profit. Instead of wholesaling 10 pounds of squash seed for $300, you may be able to retail 640 ¼-ounce packages for $2 each, more than quadrupling your gross income. You will be responsible for final cleaning, germination testing, packaging, and marketing your seed. Consider the costs of catalog production and mailing, Web page, packaging, labor to fill the packets, and shipping before you get too excited. Consider also how willing you are to serve the general public, how open you are to having your life disrupted and changed. If anything goes wrong, you are completely responsible to those who purchased your seed.

A survey of participants at the December 2004 Restoring Our Seed Conference revealed that time and land are the biggest constraints among would-be seed growers. Give careful thought to these resources before you begin. With so much to do already on your farm, are you ready to take on a new time-consuming challenge? Do you have enough land to spare so that you will be able to isolate similar crops from each other, if necessary, in order to grow seed? Can you afford to take land away from crop production to put into seed production? Or do you have idle land that would be ideal? Can you figure out a rotation that will work well both for vegetable and seed crops? Consider also your skills and interests and available equipment and tools before you launch into seeds.

You might want to start slowly, mastering the basics on a small scale or beginning with easier selfers like beans and tomatoes, before venturing into challenges like onions and leeks. Start at a modest level of commitment and work your way slowly until you reach the maximum level that is still comfortable. Don't forget to consider the possible benefits to your farm including greater diversification, the introduction of more pollinators, and possibly heightened aesthetics.

Challenges to Growing Organic Seed Crops in the Northeast

Tom Stearns of High Mowing Seeds in Wolcott, Vermont, has divided the problems of growing seed in our region into two categories: agronomic and historical. The agronomic problems are posed by our severe climate with its long, cold winters and high moisture levels.

Agronomic Challenges

The short growing season makes some long-season crops such as eggplant, peppers, and some squashes such as 'Waltham Butternut' difficult to grow in the colder years. It increases the importance of choosing varieties wisely. It limits the biennial crops that can be left in the field to a very few, forcing growers to bring most root crops and brassicas inside for the winter. It makes crop isolation by time (see chapter 3 for more details) problematic. For example, the season is almost too short and the summer heat too intense to enable more than one variety of corn to be grown per season without overlapping pollination times. On the other hand, severe winters reduce disease and insect pressures.

Summers in our region are humid with frequent rainfall, which intensifies in the fall. Often it is difficult to produce dry seed crops because they keep getting remoistened, which leads to seed-coat damage and perfect opportunities for fungal and bacterial diseases to infect and wreak havoc with the crops. For these reasons, companies like Fedco and Johnny's often farm out production for crops such as peas and lettuce to more favorable arid regions in the West. On the other hand, the Northeast is reasonably favorable for growing beans, tomatoes, cucurbits, and *Brassica* greens.

Diseases pose a particular challenge to organic growers, who are allowed fewer tools than conventional growers. Until recently, little research has been done on organic disease control. In particular, growers and regional seed companies must be aware of seed-borne diseases such as *Sclerotinia*.

Organic growers can adopt practices that alleviate some of the risks. These include wider plant spacing to increase air flow and reduce chances of disease, using rain covers over crops such as lettuce, working to improve varieties by selecting for resistance to the typical moist-climate diseases, and using acceptable post-harvest treatments such as sodium hypochlorite solution on tomato and pepper seeds.

Anthracnose (ring spot) in lettuce.

Organic growers probably also need to adopt greater isolation distances than conventional growers in order to reduce risks of cross-pollination. Although this is an area of some controversy in the reference literature, organic growers would be well advised to use the high end of the ranges cited. Anecdotal evidence is mounting (Frank Morton, Tom Stearns, pers. comm.) that pollinators are more active in organic systems, and the presence of so many more plants in flower over long periods of the season only encourages them. Even selfers such as beans and tomatoes cross more readily under organic conditions.

Organic management of soil fertility for seed crops is a largely unexplored field. Due to the seed's role in flower and fruit formation, seed crops probably require more phosphorus than market-garden crops (Stearns 2004). Too much nitrogen may cause overabundant vegetative growth, with the concomitant risk of lodging, which will encourage fungal disease. Lush, succulent growth may also retard flowering and seed development. Stearns has been cutting back on the nitrogen.

Historical Challenges

Historical problems in our area stem largely from the absence of a seed industry. When growers here have problems, it has been hard to find the answers to their questions. The appropriate seed-cleaning equipment is hard to find, there is only one public seed-testing lab in the region, and most extension agents know little about growing seed. That is beginning to change, thanks in part to aggressive efforts by organic gardening organizations such as the NOFA chapters and MOFGA. Cornell University has a program for breeding cucurbits with powdery mildew resistance and has recently been funded for two participatory farmer-breeder projects. Tom Stearns has built a full-scale seed conditioning facility in Vermont. In the person of Mark Hutton, Maine now has an extension agent with experience in the seed industry and expertise in growing seed.

Intellectual Property Rights

Plant breeders have the right to apply for legal protection of their varieties through the Plant Variety Protection Act (PVPA or PVP). PVPA gives the breeder exclusive rights to the variety for twenty years for annual crops and twenty-five years for trees and shrubs (Kloppenburg 1988, AMS USDA Web site). Breeders can license seed companies to sell their seeds and receive a royalty in return. Farmers may save seed and replant without breaking the law. Originally, farmer-to-farmer sales were permitted; in 1994 the law was amended to make this illegal (Shiva 2000). For seed production, if you have a contract, it is a seed company's responsibility to get permission to increase PVP varieties. Direct sale of these varieties without permission is illegal; they are usually indicated by a "PVP" in catalogs. After twenty years the seeds can be sold by anyone.

Genetically modified organisms, or GMOs, are not allowed under organic certification standards. They are also protected under very rigid patent laws. Companies that produce GMOs can sue for property theft if the genes cross into your crops from neighboring fields, even if you never purchased GM seed. The case of Percy Schmeiser, a canola breeder and grower in Bruno, Saskatchewan, is the best-known example of a farmer being sued for accidental gene contamination. When producing seed for sale, grow your crops as far as possible from any GM crops of the same species. With the mustard or *Brassica* family, accidental crosses between species are frequent; isolate all members of this family from GM canola.

three

Growing Seed

Pollination Biology

Selfers

"Selfers" are plants that reproduce by using their own pollen and, for the most part, actually have mechanisms that prevent effective fertilization by foreign pollen. Beans, peas, tomatoes, and lettuce are examples of self-pollinating crops. Harvesting seed from 25–50 plants should be adequate to maintain the genetic integrity of these crops each time seed is produced. These plants do not suffer from inbreeding depression because they are adapted to using their own pollen and any deleterious recessive genes have been removed from the gene pool. Fifty plants is the ideal and gives the variety enough diversity that it can adapt to different conditions, but in practice seed savers often grow many fewer plants. A word of caution: selfers can cross under certain conditions. Keep different varieties of self-pollinating crops isolated from each other, preferably by at least 50 feet. Some growers do plant different varieties, with easily identifiable characteristics, of the same crop near each other so that in the following season any crosses can be spotted and eliminated for the next growing season. This strategy may work in some cases, but dominant genes may mask unwanted crosses. Good isolation is the best solution to maintain the genetic integrity of seed crops.

'Swiss Giant' peas.

Outcrossers

Outcrossers are plants that tend to be pollinated by a separate plant or plants. Corn, brassicas, cucurbits, and alliums are classic examples of outcrossers. Cross-pollinated species are pollinated through the action of wind, insects, or other physical methods of transferring pollen from one plant to another. Outcrossing species have evolved a diverse array of mechanisms to avoid self-pollination such as self-incompatibility, dioecy (separate male and female individual plants), temporal separation (male and female flowers or flower parts mature at different times), and various flower constructs to separate the male and female parts within one flower.

Nantes carrots gone to seed in greenhouse.

At least a quarter mile or more should separate varieties of these plants. A minimum population is ideally 100 or more plants to avoid inbreeding depression. (Inbreeding depression occurs when closely related plants reproduce with each other.) Outcrossers carry what is called "genetic load"—detrimental recessive genes that decrease the vigor of a plant. These recessive genes hide in a large population where normal dominant genes are abundant and mask their bad effects. As relatives begin to pollinate each other, the same detrimental genes come together and there are fewer normal genes to mask the ill effects. Some outcrossers can self-pollinate, but this will also quickly cause inbreeding depression. Squashes can be grown in smaller populations as they are less prone to inbreeding depression than some other outcrossers. Corn and some brassicas are very susceptible to inbreeding depression. Keep populations as large as possible.

Use a few tricks to keep diversity high within your populations to avoid inbreeding depression. Include seeds from previous years with the current season's plantings because the seed from the original source should have the most diversity. You may choose to mix seed samples from different sources. Risks associated with combining samples are new seed-borne diseases, possible genetic contamination, and perhaps strain difference. In addition, growing the crop in large numbers can reinvigorate the variety.

If you are using seed for yourself and not selling it, and planting different varieties slightly closer than you should, plant in blocks and save seed from the middle of your patch. The wind or the bees are more likely to carry the pollen of another variety to the individuals on the edge while the plants in the center are likely to have been pollinated by plants within the patch (Stearns 2004).

Outcrossers commonly receive pollination two ways: by wind and insect. Wind-pollinated crops are more difficult to isolate. If you cannot isolate varieties, place bags over their flowers to prevent unwanted cross-pollination. This is very labor intensive. For a less labor-intensive method, cage or row-cover some of the insect-pollinated outcrossers. Hand pollinating squash can be enjoyable, and the seed yield worthwhile. In general, isolation is easiest; find what works for your system.

Clonal

Clonal reproduction is a third way plants can be propagated, an asexual way of producing offspring that makes identical copies with only a few rare new types produced. Potatoes, garlic, Jerusalem artichokes, ginger, and sweet potatoes are all propagated this way. There is no chance of cross-pollination. Disease, however, can build up and decrease production. Potatoes frequently accumulate viruses and "run out" after a decade or two. Many of these types of plants have changed little because they are making copies of themselves. Many of the varieties look very similar with only slight differences. Proper marking in the field and tracking during storage is critical.

Bees and Seeds

Bees, flowers, and seeds are integrally linked in the ecological web of an organic farm. Flowers connect the pollinators (bees, butterflies, flies, bats, and hummingbirds) and the pollinated (plant species), trading sugars and protein for a genetic delivery service. This exchange of genes through cross-pollination keeps open-pollinated varieties of crops healthy by continually mixing genes and reducing inbreeding. These crops then yield genetically diverse and productive seed crops.

Seed crops themselves increase the amount of pollen and nectar on a farm and provide abundant pollinator forage. Brassicas, lettuce, carrots, parsnips, and onions in most vegetable-growing operations are harvested before they flower. In seed production, these plants yield copious bee "food" that would otherwise be unavailable. Honeybees could be incorporated into seed production to provide honey as an added product.

But remember the unsung pollination heroes: our native bees, flies, wasps, and butterflies. Our native bees are especially important pollinators; among these species are bumblebees, leaf-cutter bees, and sweat bees. Honeybees, which were brought from the Old World, dominate our thoughts because of their honey and hives, which can be easily transported from field to field. Recently, honeybee populations have declined owing to colony collapse disorder (CCD).

Native bees are unaffected by CCD and can fill the pollinator gap. Even healthy honeybees may need some help from the natives. The

bumblebees and carpenter bees have a special ability called sonication or "buzz pollination" that allows them to pollinate flowers with pollen that is released only from the tips of the anthers. Anthers of this type are found in the nightshade family (eggplant, peppers, and tomatoes). Honeybees are unable to pollinate these crops. Bumblebees can be active under cooler and wetter

Bee pollinating an echinacea blossom.

conditions than honey bees. In some years they may be critical for getting a crop to fruit. Some of our New World crops have flowers that fit better with our New World bees. For example, squash flowers have to be visited only 1.1 times on average by a carpenter bee to set fruit; in comparison, they require 3.3 visits by a honeybee (Buchmann and Nabhan 1996). As organic farmers, we do not want a pollinator monoculture and should support a healthy richness of other species of bees, wasps, flies, and butterflies in addition to honeybees.

Native pollinators need native habitats and a continuous supply of nectar plants. Many of our local species nest in logs or untilled ground and feed on pollen and nectar from diverse trees, shrubs, and wildflowers.

More native plant species will include an array of larval host plants (caterpillar food) for butterflies and moths that also pollinate crops.

Be sure to leave habitat near your seed crops that fills these needs. By adding/leaving strips of untilled land with plants native to your farm, you will keep the pollinator population high all year-round, with insects at hand when your seed crop is in bloom to maximize your seed set. In providing pollinator habitat in the form of alternative foraging areas, pollen and shelter not always found in your fields, you also support beneficial insects. Thus, you may enjoy a fringe benefit of improved pest control in addition to better pollination (Altieri 1995).

Disease Control

Most seed crops are in the field longer than normal vegetable crops, increasing the amount of time that they are exposed to disease. To reduce chances for disease, start with good disease-free seed. If you are growing a seed crop, plant the best-looking most filled-out seed that does not have any fungal damage. If you suspect any seed-borne disease, use the hot-water treatment described in appendix C. Space your plants further apart than recommended for normal production. Seed-bearing plants often grow larger than those harvested for produce.

Air flow between the plants is necessary, especially in wet years, to reduce fungal and bacterial problems. For each crop listed in chapter 4, I have a recommended spacing adapted from *Vegetable Seed Production* (George 1999). Fungi and bacteria can attack the stems and fruits, decreasing production, or they may attack the seeds themselves, lowering the germination or passing disease on to the next generation.

Rotate your crops and keep them away from other close-by farms that produce the same crop as the seed you are growing to avoid spores reaching your plants. Many seed crops begin to fall over. Stake them or use other methods to keep seed and fruits from touching the ground, where many fungi reside in the soil. Finally, be vigilant. Walk your field to look for off-type plants and to watch for disease and pests. Control them as soon as possible.

Genetic Diversity

Preservation of genetic diversity is an important facet of seed saving and should be a founding principle of the movement to save heirloom crops.

Genetic diversity resides both within a variety and between varieties. There are two main strategies for conservation: one is to grow many varieties but few plants of each, the other to grow a large number of plants of one variety.

For selling seeds, the most effective approach is to grow a large population of one variety per crop species. This sort of specialization is also the best way for farmers and gardeners to conserve a variety. We need other people to conserve a large number of cultivars so that we have places to get starter seed. For production purposes, to produce the seed well, and to get a quantity of seed that can be marketed, the fewer the varieties each of us grows the better, so that the seed grower's attention is focused to get the job done correctly. I have the utmost respect for seed savers who grow hundreds to thousands of varieties, but I think most production farmers should grow fewer cultivars because the risk of seed contamination, outcrossing to different varieties, and confusion of similar types is too great. Many varieties could be maintained well and in large numbers if several farms participated in rescuing heirlooms, each doing a good job with one variety, maintaining population sizes and vigor, retaining genetic richness, proper characteristics, and production of the crop.

Conservation biologists have a rough gauge called the 50/500 rule for endangered species of plants and animals (Meffe and Carrol 1997). I believe that this rule applies to crop varieties as well. Fifty individuals in a population are sufficient for short-term survival of a variety, and most likely adequate for the long-term survival of a self-pollinated species. Five hundred individuals will most likely ensure the long-term survival of a variety, especially if the crop is an outcrossing species that requires a large population to prevent inbreeding depression. Five hundred plants almost guarantees enough seed to plant for the next season even during a "stochastic event," such as a year with drought, hail, or flood.

Plant Selection

Examine each plant carefully through all stages of crop development, from seed to harvest. Many times heirlooms are variable and occasionally may have outcrossed to other varieties. If you suspect that the variety may be inconsistent, plant more than the minimum for genetic diversity so you can eliminate off-types while keeping the population large.

The process of examining a crop and removing unwanted individuals is called *roguing*. First, you must know what your variety is supposed to look like either from descriptions, photos, other people's crops, or personal experience. Compare the crop you are currently growing to the "ideal" for that variety, to ensure your planting is true to type. Examine plant height, leaf shape, fruit color, plant habit, days to maturity, and any other characteristic that can vary. A plant that doesn't match the ideal for the variety is called an *off-type*. Remove off-types from your population. If the plant is an outcrosser, remove it, if possible, before its male flowers have opened. Some seed savers remove the off-type plant and several others around it if the male flowers have already opened, to reduce the chance of it pollinating nearby plants and thereby increasing or maintaining itself in the next generation. Many times an off-type cannot be seen until the plant has flowered and made fruit or seed, such as corn kernels of the wrong type or color. Do not save off-type kernels, and select against these types in the following generations. Eventually the unwanted trait will decrease and possibly disappear. Rogue all diseased plants. This may increase horizontal resistance to plant disease and decrease the chance that your seed will infect your next crop or someone else's crop.

Isolation

Isolation means the separation of plant varieties to ensure genetic purity. There are four main isolation techniques used to keep varieties pure: distance, timing, bagging and hand pollination, and caging techniques.

Distance

Varieties can be kept genetically true to type most simply by spatial or distance isolation. Distance is the most labor-efficient method of isolation. Outcrossing or cross-pollinated species are the most difficult to isolate using distance. Generally outcrossers need a minimum of one-quarter mile between varieties—one-half mile is preferred. More distance is advised, if possible; each crop is different, and this handbook recommends an isolation distance with each crop description (chapter 4). This distance prevents bees from cross-pollinating plants and wind-borne pollen from landing where it is not wanted. Each situation is unique. Some fields may be windy or beehives may be near, increasing the distance pollen can travel.

Self-pollinating species require only small isolation distances, 50–100 feet from other varieties of the same species, though different sources recommend different distances, and little is known about the particular details of pollination on organic farms. Organic conditions often support a high number of pollinators, and greater isolation distance may be needed than for conventional agriculture. The Public Seed Initiative handbook (2004) recommends 300 feet between many selfers.

Timing

Timing can be another labor-efficient method of isolation. Crops of the same species are planted at different times so that they are not simultaneously in flower. This method works only for plants that flower in one burst. Given a sufficiently long growing season, corn and brassicas may be time-isolated. Enough time needs to pass between plantings so that the flowering stages of the different varieties do not overlap. Timing isolation can be risky. Staggered plantings often catch up to each other if the later one gets hotter weather. Be diligent; sometimes you must cut off flowers to insure purity.

Alternate-year flowering of biennials is one more way to prevent cross-pollination of varieties within the same species. Biennials are plants that grow leaves or storage roots in their first year, then make flowers and seeds in their second. For *Brassica napus* you could plant 'Red Russian' kale during year one, over-winter it, and let it go to seed in year two. At the beginning of year two you could plant a rutabaga, and since it will be in storage mode instead of flowering mode, the varieties will not cross. Then in year three, the rutabaga will safely flower with no chance of outcrossing to the kale. During year three you could plant another *napus* or switch back to 'Red Russian' kale that would flower in year four, etc. This technique works with any biennial such as *Brassica oleracea*, beets/chard, and carrots.

Bagging and Hand Pollination

Bagging is creating a simple pollen barrier using cloth or paper bags to cover a flower or group of flowers to prevent unwanted cross-pollination. This technique is used when growing more than one variety of a species that can self-pollinate but has a high chance of cross-pollinating with another variety in close proximity. There is no need to move pollen between plants. Okra and sorghum are examples of plants with this type of reproduction. Bags are placed over the flowers when the plants are in bud and act as a pollen barrier while the flower is open. The bag can be left over flowers under most conditions until the seeds are mature. Be aware that in hot weather temperatures may get too high in the bag and cause the seeds to abort; in wet years bags can get saturated and seeds may mold.

Hand pollination is a technique using a barrier followed by manual pollination. For hand pollination the flowering structure is bagged or taped to prevent unwanted pollination. Unlike crops that can be simply bagged, some crops cannot be allowed to self-pollinate because inbreeding depression results or they are unable to produce any seed at all. Pollen has to be moved from one plant to another at the correct time. Hand pollination can be labor intensive and may not be practical for growing large amounts of seed. Hand pollination is an excellent way to increase rare stock seed that could be used as starter seed for a commercial seed crop in the future. Detailed methods are covered in each crop description.

Caging Techniques

Growers sometimes construct cages made of a wooden frame and window screen material to keep insect-pollinated plants isolated from each other. They need to be large enough to cover the desired number of plants and not have flowers touching the screen material. Flowers in contact with the screen can be pollinated from the outside and therefore contaminated with genes from another variety. This technique is good for maintaining large collections of plant varieties and could be used to increase a small sample of seed for commercial seed growing in the future. The cages are a big-time investment and take up a great deal of storage space. A sturdy floating row cover or Reemay is a quick and economical way to cover a larger number of plants. It must be fastened securely and be completely sealed at the bottom to be absolutely sure no insects can get to the flowers.

These simple caging and row-cover techniques are good for plants that can self-pollinate and do not show inbreeding depression, but are easily contaminated by pollen from other varieties. Peppers may be one crop that could be produced under cages or row cover because they can self-pollinate without major inbreeding depression, and a few plants can produce a high volume of seed.

If you are caging two varieties of a cross-pollinating species, you can cage the varieties on alternate days. The two varieties are caged or are covered with Reemay. This safely isolates them from unwanted pollen, but because the plants are self-incompatible they cannot produce seed. One cage is removed and pollinators work the flower, moving pollen from plant to plant within the variety. At night when the pollinators are not active, the row cover or cage is replaced. The cage for the other variety is then removed and the next day the plants of that variety are cross-pollinated within that cultivar. The cycle is repeated until the desired amount of seed is set; the cages then remain over both varieties until the seeds are mature. Leaving the cages prevents later-blooming flowers from being contaminated and mixing with the seed crop. This allows pollination between plants of the same kind while preventing pollinator access to the other variety. The alternate-day caging technique may also be used for species that can self-pollinate but may show inbreeding depression; this method prevents genetic contamination, at the same time allowing a proper mixing of genes within a variety.

Harvesting, Cleaning, Drying, and Storage

Seeds can be classified as wet seeds or dry seeds. Wet seeds are produced in a fruit with moist flesh, e.g., tomatoes, peppers, eggplant, and cucurbits. They are protected inside the fruit from rain, humidity, and many diseases and insects. Because of their relative imperviousness to rain, wet seeds are generally well adapted and easy to produce in the Northeast. Dry seed is much more difficult to grow in the Northeast and other regions where frequent rain and humidity occur in the later stages of seed maturation. Dry-seeded crops include the grains, sunflowers, brassicas, legumes, alliums, umbellifers (carrot, parsley, etc.), and most flowers. There is a fine balance between harvesting mature dry seeds and leaving them too long in the field, resulting in rotted seed.

One can harvest and clean seeds several ways. For dry seed, the plant has to reach the proper maturity and have low enough moisture content.

When they are ready, hang the plants in a shed or barn to finish drying. When they are completely dry, thresh them by stomping on them in a pillowcase or a tarp. Another way to thresh small amounts of dry seed is to use a tray and a board with a handle (paddle). The staff of the Public Seed Initiative's (a joint Cornell University and NOFA–NY project) mobile seed-cleaning unit recommend gluing rubber stair tread into a tray with walls. A 9-by-13-inch cake pan would work. Tread is also glued to the broad side of a board, perhaps ½-inch plywood, that would fit inside the pan. Attach a handle to the board that allows it to be picked up from the middle. Place seeds in the pan, then press the broad side of the board down on the seed heads several times until the seeds are free.

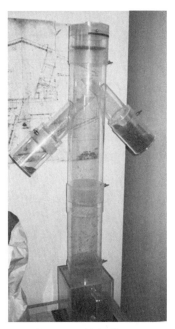

Air column seed/chaff separator.

With both threshing methods, the seeds fall to the bottom, and the dry plant material on top can be removed. After the seeds fall to the lowest point, you can run them through a series of screens made out of different sized hardware cloth. Have at least two screens stacked on top of each other so that debris larger than the seed remains in the top one, while the lower one catches the seed and the smaller debris falls out the bottom. Three screens are probably better to catch any seeds that are slightly smaller than the norm. See appendix D for a list of appropriate screen sizes for various crops. After screening, the seed needs to be winnowed—to have all the chaff blown off. For small-scale winnowing, pour the seeds from one vessel to another in front of a variable speed fan. Adjust the speed to produce air flow that is fast enough to clear the debris but slow enough that your seeds aren't blown away. Screening and winnowing could be done by hand or with a clipper mill. In chapter 4 we will go into detail with each crop and discuss larger-scale options.

Wet seed has to be extracted from the fruit. Crush tomatoes or peppers, cut open squashes. Ferment the seeds (tomato seeds in their own juice,

Once harvested, plants can hang in a shed to dry. When completely dry, thresh them by stomping on them in a pillowcase or tarp. Then run the mixture of seeds and chaff through a series of screens (shown above) to separate the unlike-sized chaff from the seed. Finally, winnow to remove the similar-sized but lighter chaff from the seed.

pepper and squash seed covered with water), then rinse until clear of debris. Once they are clean, spread them out to dry.

Drying is difficult in the Northeast. All seed needs to be completely dry before storage. Rooms with wood stoves are excellent places to prepare seed for storage. Space heaters, fans, and dehumidifiers are also excellent tools. Hit the seed with a hammer to see if it shatters, or try to snap the seed. If it shatters and snaps cleanly without bending, the seed is dry. Each type of seed has a different storage life. When completely dry, most seeds of conventional crops can also be frozen and stored for a much longer time. Glass jars are an excellent way to store seeds, but anything that is rodent-proof, grain-moth-proof, and sealed from humidity will do the job.

Germination Testing

To ensure that you have good seed, you should do a germination test. If you are selling seed, your contract will likely define minimum germination standards. If your seed does not meet the minimum standard, the company may reject it. In some cases there are penalties for falling below the minimum, and premiums for exceeding the standard. Independent seed labs such as the New York State Seed Testing Lab at the New York State Agricultural Experiment Station in Geneva, New York, will test your seed for germination rate and weed-seed contamination. Even if you are saving seed just for yourself, it is a good idea to have your seed tested. You may decide that it is not economical to plant your own seed if germination is low or if the seed is contaminated.

You can do a home germination test by counting out a known number of seeds (100 is sufficient) and planting the seeds in soil kept at the correct soil moisture and temperature for that crop. Most crops germinate within two weeks. Count the healthy seedlings and calculate the percent of germination. Also note the general vigor of the seedlings. From this knowledge you can decide which seed lots to sell or plant.

four

Details on Growing Individual Crops

Disease information, spacing, and soil pH are adapted from *Vegetable Seed Production* (George 1999). (See also *Growing Healthy Vegetable Crops* by Brian Caldwell, a companion NOFA handbook.) Not all the diseases listed occur in the Northeast, but seed purchased from other areas may bring these diseases with them. Yields for common crops are from Tom Stearns of High Mowing Seed, who has several years of seed-growing experience in the Northeast. Isolation distances are from Ashworth's *Seed to Seed* (2002) or the PSI handbook. Spacing and yield were converted from metric to U.S. customary units. The PSI manual has slightly different spacing recommendations. Adjust distances between plants and rows for cultivars of varying sizes, and change as necessary for bed production.

The *Amaranthaceae* or Amaranth Family

Amaranth *Amaranthus* spp.
 Life history: Annual
 Pollination: Outcrosser or wind
 Recommended minimum population size: 25
 Isolation: ⅛ mile or greater; bagging; can cross with weed amaranth
 Seed-borne diseases: *Alternaria amaranthi*; strawberry latent ring spot virus
 Spacing: Row 24"–27"; plant 16"–20"
 Estimated seed yield: 1,780 lbs./acre (George 1999)
 Soil pH: Not higher than 7.0
Amaranths are beautiful plants that originated in Mexico and South America. There are many named species, and at least some crossing occurs

between most types, though some crosses are rare and produce only sterile offspring. Varieties of Amaranth were banned under Spanish colonial rule in Central and South America because they were associated with Native American religious life and rituals. Amaranths have been bred for diverse uses. They are eaten as a green in Asia, Africa, and Jamaica, they are absolutely stunning ornamentals and cut flowers, and they produce high-protein seeds that can be used like a grain.

Seeds are ready when they begin to shed from the plant. Thresh by beating the whole plant on a tarp. You will create a pile of seeds mixed with chaff. Pour this material through a series of screens. The first screen should retain large material and let the seed pass. The second screen should hold the majority of seed and let smaller debris fall. You can use a third screen to catch some of the smaller-than-average seeds. After screening, only chaff similar in size to the seed remains. It will be lighter than the seed and can be blown

Amaranthus spp.

away—or winnowed. To winnow the chaff, pour the seeds and chaff from one bowl to another using the wind or a fan to blow the chaff off. You may have to transfer the seeds several times to get them completely clean. Experiment with varying the air speed and adjusting the height from which the seed is poured to retain more seed and less debris.

The *Brassicaceae* or Cabbage Family

The brassicas, formerly known as crucifers (*Cruciferae*), are in the cabbage family, which is an incredibly diverse group. Humankind through the millennia has shaped the same species into many forms for different culinary uses. Many distinct types that appear dissimilar to our eyes are biologically compatible and can produce viable offspring with each other. Members of the same species cross freely if they are grown within a half mile of each other and are flowering at the same time. Here is a list of the brassica species and the common crop types they include:

Brassica juncea are mustard greens, including the cultivars 'Green Wave' and 'Osaka Purple'.

Brassica napus includes rutabagas, Russian kales, and rape.

Brassica oleracea is a diverse crop species that includes cabbage, kale, Brussels sprouts, kohlrabi, broccoli, collards, and cauliflower.

Brassica rapa is another very diverse species that includes turnip, broccoli rabe, Chinese cabbage, Chinese mustard, komatsuna, tatsoi, mizuna, and pac choi/bok choy.

Eruca vescaria ssp. *sativa* is arugula, rocket, or roquette.

Diplotaxus tenuifolia is wild arugula or sylvetta.

Raphanus sativa, or radish, includes all red, white, pink, black, and purple root radishes including daikons, as well as pod or rattail radishes.

In general, crucifers are self-incompatible, the exceptions being *B. juncea* and *B. napus*. All self-incompatible species should have large population sizes to avoid inbreeding depression. To reduce seed-borne disease, the flowering stocks of brassicas should be staked to prevent soil fungi from infecting the seeds.

Life history: Annual or biennial

Pollination: Insects, bees, flies

Recommended minimum population size: 50–100 plants

Isolation: ½ mile; alternate-day caging; alternate-year flowering for biennial forms

Seed-borne diseases: Black rot; white blister; *Alternaria* leaf spot; black leg; black spot/wire stem; black ring spot; club root; watery soft rot; downy mildew; *Fusarium* wilt; bacteria leaf spot; turnip yellow mosaic virus

Common pests and diseases: black rot; *Rhizoctonia* stem and head rot; *Phoma* black leg; downy mildew; *Sclerotinia* blight; *Botrytis* head rot; club root; cabbage root maggot and imported cabbage worm; flea beetle

Soil pH: 5.5–7.5 (varies for individual plants; see below)

Mustard Greens *Brassica juncea*
Spacing: Row 20"–35"; plant 24"–28"
Estimated seed yield: 1,200–2,500 lbs./acre

Isolation: ½ mile

Soil pH: 6.0–6.5

Mustard greens are very easy to grow for seed. These plants are annuals. They produce large edible basal leaves. Do not harvest leaves (or do so lightly) because these will provide energy for the seeds later. The seedlings are adapted to cold weather and can be planted in March or April. Large leaves develop as the weather warms, bolting occurs, abundant flowers follow in early summer. *Brassica juncea* is considered to be two-thirds self-pollinating and one-third cross-pollinated by insects.

Windrow of *Brassica juncea*.

Brassica juncea may cross with *B. rapa* and occasionally *B. napus*; all three of these species should be isolated from each other. The flowering plants can be huge, so provide ample space: 2-foot spacing with 3-foot rows will probably be adequate—a much wider spacing than if you were producing greens for sale. Tom Stearns of High Mowing Seeds loves to grow brassicas, and he windrows large fields of mustard greens. He and his staff bend the long stalks over each other making neat "hoops" that keep the seed off the ground but allow ample air flow for the seed to dry. The seeds are ready when the bottom pods (called siliques in brassicas) are dry and begin to break open. Cut the plants at this stage and place them somewhere to finish drying. Thresh by placing the siliques on a tarp and stepping on them. Screen the seeds from the remains of the siliques, then winnow the chaff by pouring the seeds back and forth between buckets or bowls. You could also use a belt thresher or stationary thresher and an air column for winnowing. Combines can both harvest and thresh. Further cleaning can be done with a clipper mill with the appropriate-sized screens. Brassica seeds sometimes break in threshing or are withered and do not germinate. The filled-out round seed will germinate the best. The other seed can be separated out using velvet rollers or a spiral seed separator. Round seeds roll faster and spill over the sides of the spiral while the shriveled or broken seed stay on it and then fall off at the bottom into a separate container.

Russian Kale and Rutabaga *Brassica napus*
Spacing: Row 20"–35"; plant 4"–8"
Estimated seed yield: 1,300 lbs./acre
Soil pH: 5.5–6.8

Brassica napus, like the other members of the mustard family, is an easy species to grow for seed. It is slightly more difficult than mustard greens because it is a biennial. This species will not bolt unless it is exposed to cold weather, a process called *vernalization*. 'Red Russian', 'Wild Russian', and 'White Russian' kales are vigorous and grow well. These varieties survive the winter in Connecticut and are the only crucifers I would recommend overwintering outside in the Northeast. Some individuals winter-kill, but losses should be minimal in most locations in the Northeast if plants are mulched and placed under row cover. Test a small number for winter hardiness before trying to grow a large seed crop. These kales grow wonderfully luxurious leaves their first year, which eventually winter-kill, but the stem survives and resprouts the second year, sending up flowers that are worked by the bees. Separate varieties of *B. napus* by a half mile or more and also isolate them from other *Brassica* species if possible, especially *B. rapa*. Alternate-year flowering of this biennial is one more way to prevent cross-pollination of varieties. The siliques develop in a similar fashion to mustard greens and are harvested at the same stage of development. Threshing and seed-cleaning techniques are similar to *B. juncea*.

Cabbage, Kale, Brussels Sprouts, Kohlrabi, Broccoli, Collards, and Cauliflower *Brassica oleracea*

Cabbage
Spacing: Row 24"; plant 12"–24"
Estimated seed yield: 620 lbs./acre
Soil pH: 6.0–6.5

Cauliflower
Spacing: Row 24"; plant 12"–24"
Estimated seed yield: 360 lbs./acre
Soil pH: 6.0–6.5

Brussels Sprouts
Spacing: Row 24"; plant 12"–24"
Estimated seed yield: 530 lbs./acre
Soil pH: 6.0–6.5

Kohlrabi
Spacing: Row 24"; plant 12"–24"
Estimated seed yield: 620 lbs./acre
Soil pH: 6.0–6.5

Brassica oleracea is the most commonly cultivated brassica species. Its ancestor is a wild cabbage that still grows along the seashore throughout Europe. All the forms we know today are products of human selection over thousands of years. All forms intercross if flowering near each other at the same time. Isolation is necessary to produce pure seed crops. The use of alternate-year growing techniques and caging are effective ways of preventing unwanted crosses. *Brassica oleracea* are cross-pollinated plants and must be hand or insect pollinated. Many members of this species are self-incompatible and show inbreeding depression, therefore it's important to grow large populations. These plants are biennials but are not hardy in our climate. If left in the field through the winter, they often die. Generally, in the Northeast, all members of this species are grown one season and potted up and

'Savoy' cabbage.

taken into a root cellar for storage. Check the moisture level to prevent them from drying out. In late March or April set the plants back out in the garden. Cut an X across the top of the cabbage heads to allow them to bolt more cleanly (Johnston 1983).

Broccoli and cauliflower can be grown as annuals if started inside in February or early March and set out in April. The plants begin to bolt with the heat of the summer. Seed cleaning is similar to other brassica species. Because of seed-disease problems and the labor involved in overwintering *B. oleracea*, many of the Northeast seed companies purchase seeds from other areas of the country where the climate is suited to seed production.

Turnip, Broccoli Rabe, Chinese Cabbage, Chinese Mustard, Komatsuna, Tatsoi, Mizuna, and Pac Choi/Bok Choy *Brassica rapa*

Spacing: Row 20"–35"; plant 4"–8"
Estimated seed yield: 1,200–2,400 lbs./acre
Soil pH: 6.0–7.5

This species is becoming extremely important in the Northeast as a main ingredient of mesclun mixes. These Asian greens grow well in our climate but are prone to flea beetle damage. This species is an annual but usually has to have enough cold to trigger flowering. Tatsoi bolts and flowers very quickly and is one of the easiest plants to grow for seed. Mizuna and turnips are a bit trickier and should be planted in late March or early April to get enough cold weather to trigger flowering. Isolation, pollination, and seed cleaning methods are similar to the rest of the *Brassica*.

Radish *Raphanus sativa*

Spacing: Row 20"–35"; plant 2"–6"
Estimated seed yield: 800–1,000 lbs./acre
Seed-borne diseases: Gray leaf spot; black leaf spot; leaf spot; anthracnose/leaf spot; root and stem rot; black leg; damping-off/canker; bacterial spot; radish yellow edge virus; tobacco streak virus; turnip mosaic virus
Common pests and diseases: Flea beetle; cabbage maggot
Soil pH: 5.5–6.8

Radishes are in a different genus but belong to the same family as the brassicas, and many of the growing requirements and pollination and seed-cleaning techniques are similar. Red round radishes are quite humdrum when compared to the other wonderful varieties that exist. There are black, white, and beautiful purple-skinned types, as well as the daikons, which grow 20 inches long and remain mild and crisp. Additionally, pod or rattailed radishes, which have extremely long seed pods (up to 12 inches), are used to spice up stir fries or eaten raw or pickled. All types cross.

Most radishes are self-incompatible and need to be cross-pollinated by insects. Recommended isolation distance is a half mile. Wild radishes (*Raphanus raphanistrum*) will also cross with cultivated radishes. I have had several white, skinny, fibrous, pungent radishes show up in some of my seed crops, and this weed is the likely source of those traits. Seed can be grown seed-to-seed (the seeds are planted, the roots grow and the plants bolt in place) or seed-root-seed. In the latter method, the roots are pulled before bolting for inspection or vernalization—cold treatment— after which they are placed back in the ground to bolt and go to seed. I prefer to use the seed-root-seed method to examine and rogue unwanted roots. After inspection, place remaining roots in a refrigerator for two to three days of cold treatment. This makes them flower all at the same time, insuring that many plants cross-pollinate for genetic diversity. As with other brassicas, radish flower stalks should be staked or supported in some way to keep pods from touching the soil.

Seeds are mature when seed pods turn light brown. They do not open by themselves, unlike the other brassica species. Threshing is always a challenge. A knobby rolling pin used in conjunction with a mat of stair tread in a tray works well to break open seed pods. The pods are rolled over and broken but the seeds can slip into the grooves and be unharmed. Or place whole seed stalks in tubs and step on them until the pods break open and the seeds are free. Winnowing is the same as for other brassicas.

Arugula, Rocket, Roquette (*Eruca vescaria* subsp. *sativa*), and Wild Arugula or Sylvetta *Diplotaxus tenuifolia*

Spacing: Row 20"–35"; plant 2"–6"
Estimated seed yield: 1,200 lbs./acre
Soil pH: 6.0–6.8

The Romans grew this ancient crop. Currently, it is considered a gourmet salad green like mizuna and tatsoi and is becoming increasingly important to growers in the Northeast. Few named varieties are available: 'Astro' in the United States and 'Apollo' in Great Britain. Wild arugula is a different genus and species and so will not cross with the common arugula. It is smaller and more bolt resistant. Arugula is annual and can take some frost. I have seen it thriving in Vermont at High Mowing Seeds, so it is likely to mature anywhere in the Northeast. Arugula is self-incompatible and needs to be cross-pollinated. It is unlikely you will have contamination problems because so few varieties are offered in the United States, but if you have a named type and you want it to be pure, isolate or cage it similar to the other brassicas. Seed harvest and cleaning is similar to *Brassica rapa*.

Note: Arugula seed cannot be hot-water treated easily (see appendix C) to reduce seed-borne disease. If treated, seeds need to be planted immediately while wet, or dried quickly in some sort of tumble drier.

The *Chenopodiaceae* or Beet Family

Beets and Chard *Beta vulgaris*
Life history: Biennial
Pollination: Wind
Recommended minimum population size: 25
Isolation: Bagging; 1 mile or greater
Seed-borne diseases: Seedling rot/leaf spot; *Alternaria* leaf spot; *Colletotrichum dematium f. spinaciae*; *Cercospora* leaf spot; powdery mildew; *Erisiphe betae*; *Fusarium* sp.; downy mildew; black leg/damping-off; *Ramularia* leaf spot; silvering of red beet; bacterial blight; *Arabis* mosaic virus/beet ring spot; tomato black ring virus; raspberry ring spot virus; *Lychnis* ring spot virus; eelworm canker; *Phoma* leaf spot
Common pests and diseases: *Cercospora* leaf spot
Spacing: Row 35"; plant 12"–18"
Estimated seed yield: 1,500–2,500 lbs./acre
Soil pH: 6.0–6.8

Beta vulgaris is originally native to the Mediterranean and includes both beets and chard. One was bred for the use of its leaves, the other for its delicious storage roots. Beets include varieties grown for the table and types called mangels, which were used traditionally as animal fodder. Mangels are rapidly disappearing, and some varieties may be in danger of extinction. The color variation available in seed catalogs for all these crops has blossomed into a rainbow in recent years. There are golden, white, and striped beets, and there are orange, white, yellow, and red-stemmed chard.

Beets, chard, and mangels are wind-pollinated biennials; they will cross with each other. Isolation distances of over a mile are recommended; bagging is possible, but I think not worthwhile. Barriers of tall border plants such as corn, sorghum, hemp plants (in Europe) or trees can be used to block the wind and reduce the isolation distance between varieties. In our region, overwintering in the field is unreliable. Dig up beets and mangels, inspect and select them and store them in a cool, moist cellar. Transplant Swiss chard plants into pots or bins and store them for the winter. Clip the leaves of beets, chard, and mangels to prevent the roots from drying out. Cut the leaves about 1 inch above the crown where the young leaves are emerging. Swiss chard may overwinter with mulch. After being transplanted out for the second year, the plants will bolt and produce many small

Beet.

wind-pollinated flowers. When the seeds turn light brown they are ready for harvest. Thresh the seeds by stomping on them on a tarp or in a bin. Screen and winnow them to remove the chaff. Many beets produce clusters of three seeds; these should be left intact.

The *Compositae*

Lettuce *Latuca sativa*
Life history: Annual
Pollination: Insects, bees
Recommended minimum population size: 25
Isolation: Caging; 50–100 feet
Seed-borne diseases: *Alternaria dauci*; ring spot/anthracnose; *Pleospora* leaf spot; drop/watery soft rot; southern blight; *Septoria* leaf spot; leaf blight; lettuce mosaic virus; lettuce yellow mosaic; tobacco ring spot virus; tomato black ring virus (ring spot strain)
Common pests and diseases: Slugs; downy mildew; *Rhizoctonia* bottom rot; *Sclerotinia* drop; *Botrytis* gray mold
Spacing: Row 20"–24"; plant 8"–12"
Estimated seed yield: 200–900 lbs./acre
Soil pH: At least 6.0, preferred 6.5

Lettuce has been cultivated since the ancient Greek and Roman eras. Its ancestral species is prickly lettuce (*Latuca serriola*) which is an occasional weed in the Northeast. This fairly uniform wild plant gave rise to a stunning and mind-boggling array of forms. There is even an Asian type called *celtuce* that is grown for its edible, celery-like stem.

Most lettuce will mature in the Northeast; there is a vast choice of varieties. One of our local Connecticut greens growers, Bryan O'Hara of Tobacco Road Farm, is a big fan of 'Red Sails'. Bryan also recommends 'Anuenue', developed at the University of Hawaii for summer planting.

Lettuce bolting.

For late fall/winter growing under row cover he suggests 'Winter Density' and 'Winter Marvel'. He also has a farmer-selected red oak-leaf type that he received from California that resists bolting. There is a lot of room in lettuces for farmer improvement, and Frank Morton of Wild Garden Seeds has stepped in and bred some wonderful new varieties. Note that some lettuces will have thermal dormancy and will not sprout in the heat of summer, so to get a seed crop it is best to plant in the spring. Lettuce is prone to many diseases in the humid climate of the Northeast. At Turtle Tree Seed, they construct small lettuce "huts" of corrugated clear plastic supported with metal posts that keep rain off the plants and then use drip irrigation. This method of cultivation dramatically increases seed production and reduces seed-borne disease.

Lettuce is self-pollinating, and a 50-foot isolation distance is recommended between varieties. Heading-type lettuce should be slit across the top to allow the flower stalk to emerge; other types bolt fine on their own. Small yellow flowers appear, and seeds are mature twelve to twenty-four days after flowering.

The *Cucurbitaceae* or Squash Family

The cucurbits include summer and winter squashes, melons, cucumbers, watermelons, and hard-shelled gourds. The family is tropical but has been adapted to our region over the millennia.

Squash and Pumpkin Characteristics
Life history: Annual
Pollination: Insect, bee
Recommended minimum population size: 25
Isolation: ½ mile or greater; hand pollination
Seed-borne diseases: Angular leaf spot; scab; *Didymella* leaf spot; bacterial leaf spot; bacterial fruit blotch; *Fusarium* wilt; *Alternaria* leaf spot; anthracnose; gummy stem blight/black rot; *Septoria* leaf spot; cucumber mosaic virus; muskmelon virus; squash mosaic virus.
Common pests and diseases: Powdery mildew; black rot; *Phytophthora* blight; angular leaf spot; bacterial wilt; several viruses; squash vine borer; striped cucumber beetle; squash bug; spotted cucumber beetle
Spacing: Row 3' (bush) 10'–12' (vine); plant 3' (bush) 10'–12' (vine)
Estimated seed yield: 300–900 lbs./acre, highly variable; high nitrogen can increase yield
Soil pH: 5.5–6.8

There are five domesticated species of squash: *Cucurbita pepo*, *C. maxima*, *C. moschata*, *C. argyrosperma* (aka *mixta*) and *C. ficifolia*. All of these species originated in the Americas. In general, *C. ficifolia* is not grown in our area and will not be covered. The different species can be identified using leaf, pedicel (stem of the fruit), fruit, and seed characteristics (Simpson and Ogorzaly 2001).

Cucurbita pepo leaves are rough and generally pointed at tips of leaf lobes; the pedicel (stem of the fruit) is five-angled, hard, retaining the green color in storage or bleaching out to a "straw color"; the fruits can be orange, yellow, white, green, or striped. Seeds are light beige (or green in hull-less types) with a smooth margin.

Cucurbita moschata has fuzzy leaves with rounded lobes; the pedicel is five-angled, hard (often flared at the base), drying to a golden color; the fruits can be tan, orange-tan, green, or mottled green and tan; seeds are beige or rarely dark tan with a pitted seed coat and a scalloped margin.

Cucurbita maxima leaves are rough but rounded; the pedicel is soft and spongy when dry; the fruits can be green, blue, white, pink, or orange, and can be extremely large; seeds are beige or tan, generally very thick with a smooth margin.

Cucurbita mixta has pointed lobed leaves; the pedicel is semi-spongy with rounded angles and can be very enlarged; the fruits are generally white, green, or white-and-green striped; seeds are light beige, and the flat sides can have a cracked appearance; the margins of the seeds are scalloped, and in some varieties are extremely large and silver.

Squash and Pollination

These species generally will not cross-pollinate with each other but the possibility cannot be ruled out. I have seen an occasional cross between *C. maxima* and *C. moschata* but the resulting plants have been sterile; this cross is also mentioned by Johnston (1983). *Cucurbita mixta* and *C. moschata* are also reported to cross occasionally (Merrick 1990 and Goldman 2004) with a high percent seed set when *C. mixta* is the maternal parent; these hybrids are fertile. *Cucurbita pepo* will not cross with *C. maxima* or *C. moschata* under normal conditions. I have intentionally attempted to cross *pepo* and *moschata* many times without any resulting seeds.

Male (*left*) and female squash blossoms.

Squashes and pumpkins are monoecious, meaning they have male and female reproductive structures on the same plant, but in different flowers (see accompanying photo). They are insect-pollinated with bumblebees, squash bees, carpenter bees, and honeybees as the most important pollinators. Isolation is the easiest way to get a good seed crop. Cucurbits are not especially prone to inbreeding depression, but the more diverse the open-pollinated population, the more likely there will be a good crop.

Hand pollination with the squashes is easy and rewarding if you have the time and patience. A seed crop of about 5 pounds could easily be produced that way. Hand pollinating enables you to grow several varieties of the same species in close proximity and produce multiple small seed crops. If hand pollinating, give yourself ample space so that vines do not grow into one another and you don't have to waste time following them to find the correct flowers. Cucurbits have either bush, semi-bush, or vine plant habits. Bush plants are compact with very little lateral growth (generally 3–4 feet in diameter), semi-bush generally stay compact early in the year and spread sending out short vines as they grow, and vines essentially have only lateral growth, some reaching 20 feet in length. Bush varieties are easy to work with because their compact nature allows you to grow more plants in a smaller space, keeping your genetic diversity high. Short vines are neat and do not ramble into other varieties. If you have the space, vining types are also fine to work with.

Hand pollination of squashes begins when the flower buds have grown to full size and have turned bright orange. Inspect the plant in the evening when it is most obvious what flowers will open the next day. Male flowers have longer stems; female flowers have short stems with a miniature squash under the flower. Tape the orange flower buds around the top with masking tape or close them with a bread bag twist tie. Taping prevents pollination by bees carrying pollen from another variety. Ideally, pollinate each female flower with male flowers from three or more plants for genetic diversity. The next morning the flowers will have puffed out if they are ready. (If taped too soon they will not be enlarged.) Clip the male flowers with a short section of stem and bring them to the plant that will act as the female parent. Remove tape and peel the petals off the male flowers, remove tape from the female flower but leave the petals. Use the anther to "paint" the pollen of the males onto the stigma of the female flower. Close and securely shut the petals of the female flower so no insect can enter. Watch out for bees: they often zip in when you are trying to pollinate and can contaminate the flower with unwanted pollen. The female flower should be marked after pollination. I use orange survey tape and write the date pollinated on it with a permanent marker. Sixty days after pollination is adequate for the seeds to mature within the fruit.

The fruits are mature when the rind is hard and cannot be easily injured with a fingernail. Most varieties will turn orange or develop some sort of orange blush when ready for seed harvest. Pink, blue, and bright yellow squashes may be the exceptions. For *C. maxima* varieties the stem will be completely dry and brown at full maturity. If the maturity of the fruit and seeds is in question, leave the fruit until a later date. It is generally beneficial to leave the seeds in the fruit for longer than sixty days because the seeds increase in viability as the fruits go through an after-ripening process. The only dangers to the seeds are that a fruit will rot and fungus or bacteria will damage the seeds. If left for several months, some varieties have seeds that sprout within the fruit; when removed and dried, these sprouts die. Premature sprouting often occurs in *C. pepo* summer squashes, some *C. moschata* types, and an occasional *C. maxima* variety after four months in storage.

Squash and Pumpkin Species

PEPO

Cucurbita pepo is the most commonly grown species of squash. *Cucurbita pepo* is interesting because it was domesticated twice: once in Mexico, and then again in the eastern United States, making it one of the few domestications that has occurred in this region (Smith 1995). This species includes jack-o'-lantern pumpkins, acorn, 'Spaghetti', zucchini, straight-neck, crookneck, 'Sweet Dumpling', 'Delicata', and pattypan squashes, also small ornamental gourds such as 'Crown of Thorns', 'Bi-Colored Spoon', and 'Jack-Be-Little'. Some ornamental gourds and all wild gourds (which are weeds in the Midwest and southern United States) contain *cucurbitacin*, an intensely bitter substance. These gourds also belong to *C. pepo* and freely hybridize with cultivated squashes of this species. The bitterness is dominant, and the crop can turn out bitter the next season after pollination has occurred. If edible squashes are pollinated by a gourd, that fruit is not affected, but the seeds within that fruit when planted will produce intensely bitter squash/gourd crosses. High levels of cucurbitacins are poisonous to humans. Be sure to keep your edible squashes as far as possible from any *C. pepo* gourds.

The zucchini and pumpkin types with large stems and orange or green rinds represent the Mexican domestication. The smaller-stemmed, fluted, striped, yellow, or white squashes seem to have come from wild gourds native to the Ozarks and the Mississippi River Valley, including acorn, pattypan, yellow summer squash, and the ornamental gourds. All varieties from both domestications are cross-compatible and should not be grown for seed within a half mile of each other, further if possible. In general, varieties of *C. pepo* mature well in New England, and I have never had a failure in Connecticut

Cucurbita pepo **diversity.**

when planted on time with proper fertilization and pest control. Follow general pollination and seed-saving instructions for all squash.

MOSCHATA

The butternut squash is the best known *Cucurbita moschata*. The ancestor of the butternut types is still unknown, but most likely came from Columbia (Goldman 2004). These squashes are tan, green, or mottled tan and green. Most of the Central American, southwestern U.S., and southern U.S. types may take 120 days or more to mature. Many of these are a gamble to grow in New England, but I have been successful with some, and they are often more interesting in appearance and sometimes sweeter than northern butternut types such as 'Waltham Butternut'.

If you want to grow anything nonstandard for seed, trial many varieties. 'Waltham Butternut', 'Ponca', 'Burpee Butterbush' (one of the few bush types), and 'Long Island Cheese' (milk) pumpkin generally mature in our climate. 'Canada Crookneck' is a relatively unknown heirloom *C. moschata* offered by Eastern Native Seed Conservancy that matured well for them in the Massachusetts Berkshires. It is a butternut type with a curved neck, and the flesh is deep orange and sweet.

C. moschata squashes are resistant to insects and diseases and have a long storage life. They are especially noted for their ability to withstand squash vine borer. Some *C. moschata* may be day-length sensitive, meaning they will not flower until days begin to shorten. I have tried growing *Cucurbita okeechobeenensis* or the Okechobee gourd, a day-length-sensitive, tropical species that is the source of powdery mildew resistance in our cultivated squashes. These plants grew huge vines but did not flower until late September and did not produce seed.

Maxima

Cucurbita maxima is most commonly represented by the buttercup and hubbard squashes. The wild version of these squashes is *Cucurbita andreana*, a gourd from South America. This species varies greatly; there are pink, green, blue, and orange types. Included in *C. maxima* are 'Red Kuri', 'Gold Nugget', all buttercup types, giant pumpkins such as 'Prizewinner' and 'Dill's Atlantic Giant' and 'Jumbo Pink Banana'. They all have a fleshy, round green stem that dries down into a tan corky mass when mature. Many of these squashes are very sweet and have good texture. Most varieties of this type will mature in New England. *Cucurbita maxima* squashes are very prone to insect damage, with striped cucumber beetle and squash vine borer especially severe problems. This species is fairly resistant to powdery mildew. Both bush and vine types exist.

Squash and Pumpkin Seed Cleaning

Seeds can be extracted in the field, or the fruits harvested and placed in storage where the seeds will continue to mature. Seeds can be extracted by hand when you cut the squash in half to cook. If processing a large amount of winter squash, you may want to use the squash meat for canning, freezing, or livestock. Mechanical seed extractors that crush the fruit and separate the seed from the pulp would be appropriate for seed crops of 100 pounds (which would be a couple thousand pounds of squash). High Mowing Seeds currently rents out such a machine and may have some available for sale in the future.

Wash the seed and separate out the strings. For small amounts use a colander, for larger amounts I use a ⅛-inch hardware-cloth screen mounted on a 2-by-4-foot wooden frame. The strings and leftover flesh fall through the screen, leaving the seeds. Fermenting first makes the unwanted debris softer and more likely to screen out. Light fermentation—using just enough water to cover the seeds—also reduces seed-borne disease. Fermenting should be done only for twenty-four to forty-eight hours. If allowed to continue too long fermentation causes the seeds to look unsightly and decreases germination. After fermentation, place seed on screens to dry. Near a wood stove is a good environment, but be sure the seeds do not get above 95°F. High Mowing Seeds has a simple plywood tunnel about 8 feet long in which they place seeds, then run a box fan at one end in a room with space heaters and dehumidifiers. After the seed is dry it may need a light rubbing to remove extra bits of squash. It's then ready for winnowing. Squash seeds will store for six years under conditions of low humidity and cool temperatures.

Melons *Cucumis melo*

Cucumis melo includes cantaloupe, Armenian cucumber, Crenshaw, honeydew, pocket melons, and vine peach types. This species seems to have been domesticated in West Africa, and wild types occur through the Old World tropics. Introduced weedy or dudaim types are found in crop fields of the southeastern United States and can cross freely with cultivated melons.

Pollination: Insect, bee; hand pollination very difficult

Isolation, seed-borne diseases, and major pests: Similar to
  squashes

Spacing: Row 5'–6½'; plant 1'–3'

Estimated seed yield: 150–200 lbs./acre

Soil pH: 6.0–6.8

There are four major types of melons:

> *cantaloupensis*—true cantaloupe, which are medium-sized,
>   warty or scaly, commonly grown in Europe
> *reticulatus*—most common in the United States; includes
>   "muskmelon" and "cantaloupe"
> *inodorous*—"lacks odor" includes Casaba, Crenshaw,
>   Christmas, canary, honeydew
> *flexuosus*—long types used like cucumber, including
>   Armenian cucumbers

Isolation is the only practical way to grow these seeds. One other possibility is to grow melons in a greenhouse that is completely screened off to the outside and has a bumblebee hive inside. Hand pollination is not commercially viable; approximately 80 percent of the hand crosses do not set. Armenian cucumbers, despite their name, are really melons that have been selected for a cucumber-type use. They have excellent eating qualities when compared to a *C. sativa* cucumber. Armenian cucumbers should

'Prescott Fond Blanc' melon.

not be grown with melons because they will cross and ruin your seed crop. Standard cucumbers do not cross with melons, or make them bitter. Melons have variable sex expression. Some varieties are andromonoecious, having hermaphroditic flowers containing male and female organs as well as separate male flowers. Other varieties are just monoecious with different male and female flowers on the same plant.

'Minnesota Midget', 'Golden Champlain', and 'Ralph Dyer's' melons are easy varieties to grow. C. R. Lawn uses 'Golden Gopher', which will mature in Maine, as his measuring stick for flavor. Melons are generally longer, hotter season crops, but several mature in short-season areas. Most heirloom melons are not as sugary as modern varieties. Peter's Seeds and Research (PSR) offers many gene-pool varieties that can be adapted to your local conditions. Cornell University has bred melons for disease resistance such as the powdery mildew resistant version of 'Delicious 51'. The Seed Savers Exchange offers several eastern European and old Soviet heirloom melons.

When melons are ready to eat, the seeds are mature. A melon is ripe when it generates a wonderful melon aroma. As they ripen, melons change color and develop a yellowish or orange background hue. Melons in the *reticulatus* group "slip" or detach from the vine when ripe. Most growers can extract the seed by hand and pull off any of the extra flesh attached to the seeds. A twenty-four-hour fermentation is the common method to extract high-quality seed. Fermentation also helps reduce seed-borne disease. If fermented too long, melon seeds sprout and then die when dried. Decant debris and wash seeds in a colander or on a screen, then spread out to dry.

Cucumbers *Cucumis sativus*
Pollination: Hand pollination difficult
Isolation, seed-borne diseases, and major pests: Similar to
squashes

Spacing: Row 80"; plant 4"–8"
Estimated seed yield: 300–500 lbs.
Soil pH: 6.5 or slightly higher

This crop seems to be descended from wild plants of the Himalayas. *C. sativus* will not cross with any other species. Cucumbers have more diversity than most gardeners realize. There are white ones, yellow ones, round ones, and one called 'Poona Kheera' that looks like a fingerling potato. Additionally the plants can be short vine "bush types" or have long vines.

Isolation is the only practical way of producing a seed crop. Hand pollination is possible, and a much higher percentage of fruits set than in melons, but it is tedious for the amount of seed produced. Many varieties will grow in our climate. I am partial to the odd ones and have grown lemon cucumbers for years. In addition, my wife and I have been successful in growing 'Early Russian' cucumbers, 'Beit Alpha', and 'Poona Kheera'. Lemon cucumbers seem to grow well even with cucumber beetle pressure. Striped cucumber beetles are attracted to the bitter substance cucurbitacins. Because the beetles crave these awful tasting chemicals, bitter cucumbers get the most insect damage. You can select against insect-attracting plants and bitter fruits at the same time. If the cucumber is going to be bitter, so will the first leaves of the seedling, called the cotyledons. Take a small slice of a seed leaf and taste it. If it is bitter, discard the seedling.

Leave fruits grown for seed on the plant much longer than cucumbers for eating. Allow the fruit to get oversized and turn dark yellow. They generally can keep for some time, so if in doubt leave them until just before frost. Cut the fruits in half and scoop the seeds out. Seeds are surrounded by a gel capsule that cannot be washed off and needs fermentation to be cleaned. Put the gel and seeds into a five-gallon bucket and allow them to ferment for a few days. The contents of the bucket will bubble and froth. Stir the seeds, and when it seems the seeds are free of the capsule and are sinking to the bottom, pour the debris off the top. As the seeds sink, decant the unwanted pulp several times, then wash the seeds in a colander or on screens and spread out to dry. A small seed crop of about 5 pounds can easily be processed in this way.

There are several disease-resistant varieties on the market. The most famous and widespread are the 'Marketmore' series: 'Marketmore 76', '80', '86', and '97'.

Horned Melon, Jelly Melon, Kiwano *Cucumis metuliferous*

This odd and obscure crop is seldom grown in short-season areas, but can mature in Zone 5 and has market potential as an interesting ornamental. It looks like a spiny, short, bright orange cucumber. It is a warm-season crop like its cousins the cucumber and melon. It is possible to produce horned melons in Connecticut if started indoors around the same time you would start tomatoes or possibly watermelon. Plant them in the garden at the same time as other melons. Given abundant fertility they will produce dozens of fruits per hill. The flavor is lacking, but breeders are working to increase sugar content in this crop. The fruit can keep for six months. The seeds are extracted and processed in the same manner as cucumbers.

Watermelons *Citrullus lanatus*

Pollination: Insects; hand pollination

Seed-borne diseases: *Colletotrichum* sp.; gummy stem light/black rot; *Fusarium* wilt; anthracnose; bacterial fruit blotch; *Pseudomonas pseudoalcaligenes* ssp. *citrulli*; squash mosaic virus

Spacing: Row 47"–71"; plant 35"–47"

Estimated seed yield: 100–200 lbs./acre

Soil pH: 5.0–6.8

Watermelons are delicious garden treats that originated in Africa and spread throughout the Old World by trade. Watermelons were brought to the Americas and quickly adopted by Native Americans.

Pollination is by insects. Hand pollination is similar to the squashes but the flowers are much smaller. Some watermelon varieties differ in having perfect flowers with male and female organs in the same flower. Isolation is the best bet for getting a commercial quantity of seed. Several varieties will mature in New England. I have grown 'Sugar Baby', 'Cream of Saskatchewan', 'Amish Moon and Stars', and, one hot year, 'Cherokee Moon and Stars' and 'New Hampshire Golden Midget'. I have heard of other growers producing 'Sweet Siberian' and 'Early Moonbeam' in our region. 'Orangeglo' and 'Verona', two varieties with excellent flavor, have been successfully grown in central Maine.

Seeds are fully mature when the melons are about ten days past optimum stage for fresh-market harvest. One Connecticut gardener, Dave Crocker, looks for four things to tell if a watermelon is mature: (1) tap it and listen for the dull thud, or *pink, pank, punk*, with *punk* being the ripe sound; (2) see if the tendril where the fruit attaches to the vine is withered and brown; (3) the fruit has a yellow spot on its underside; and (4) the fruit becomes slightly ribbed. Watermelons do keep. Underripe fruit may last until late December at room temperature. The seeds keep maturing in storage. Maturing seed in storage is a last ditch measure and may not yield the most viable seed. It should be used only to keep stock seed going in case of a crop failure. Extraction by hand can be difficult, but you can invite friends over for watermelon and ask them to spit into bowls. I have cut the flesh out with a knife and placed it in bowls or buckets. It can just be squeezed by hand or mashed with a potato masher until all the seeds have come free. Then add water. Most of the good seed sinks to the bottom and the pulp and bad seeds can be poured off. Wash the seeds in a colander or on a screen and spread out to dry.

Luffa *Luffa aegyptiaca* and Hard-Shelled Gourds *Lagenaria siceraria*
Pollination, spacing, and isolation: Similar to other cucurbits
Soil pH: 6.0–6.8

Hard-shelled gourds and luffas require a long growing season, but may be grown for seed in New England if started early inside or in a greenhouse. Plant the seed at the same time as you would watermelons, and transplant them into the field at the same time as other warm-season crops when all danger of frost has passed.

Hard-shelled gourds (*Lagenaria siceraria*) are native to most tropical areas of the world. These are not the striped, warted, and colored gourds used in the fall for decoration. Hard-shelled gourds are mainly used for containers and secondarily for ornamentation, but a few Italian and Asian varieties are eaten in the immature stage in a similar manner to a summer squash. Varieties include the 'Maranka' or Dinosaur gourd or Caveman's club, bottle gourds, 'Bird House' gourd, 'Dipper', 'Tobacco Box', and the famous Papua New Guinea 'Penis Sheath' gourd. More information is available from the American Gourd Society on varieties, growing, and uses.

Lagenaria has an odd pollination method. The white flowers open in the evening and are pollinated by moths, though bees do occasionally visit. I have had poor fruit set in the past and resorted to hand pollination. The fruits should be left on the vine as long as possible and then brought into a warm area with good air circulation. When the fruit is hard and dry, cut your way in or use a hammer to remove the seeds. The dust is said to be irritating if inhaled; clean seed wearing a dust mask or in a well-ventilated area. Separate the seeds and the pulp by hand or use a ¼-inch screen.

Luffa or vegetable sponge is another tropical member of the *Cucurbitaceae* that is mostly used as a good scrubber in the shower. Luffa has to be started indoors because of its long season. There are two species, *Luffa aegyptiaca*, the smooth luffa, and *Luffa acutangula*, the angled luffa. The flowers are large, yellow, and insect-pollinated. Isolation is generally not a problem because there are few varieties and very few people grow them. In Asia, the fruits are eaten when young, and luffa is sometimes called Chinese okra. Leave the fruits on the vines as long as possible but bring them in before frost. Cut off the ends of the fruits and submerge them in water. The seeds will loosen as the fruits ferment and can be squeezed out. The seeds will look like black pumpkin seeds if mature. Spread them out to dry. The leatherlike rinds are also removable after fermentation, then the spongelike fruits are hung to dry to be used for scrubbing.

The *Gramineae* or Grass Family

Corn *Zea mays*

Life history: Annual

Pollination: Outcrosser; wind

Recommended minimum population size: 100

Isolation: ½ mile or greater; hand pollination

Seed-borne diseases: Kernel rot; charred ear mold; southern leaf spot; anthracnose; smut; dry ear rot; gibberella ear rot; *Fusarium* spp.; seedling blight; cob rot; seed rot/blue-eye; crazy top; white ear rot; flase or green smut; *Ustilago maydis*; Stewart's wilt; maize leaf spot virus; maize mosaic virus; sugar cane mosaic

Common pests and diseases: Stewart's wilt, smut, rust, corn borer, ear worm

Spacing: Row 28"–39"; plant spacing varies depending on cultivar size

Estimated seed yield: 1,000–1,500 lbs./acre for sweet corn, up to 3,000 lbs./acre for flint or dent types

Soil pH: 5.5–6.8

Corn is one of the three sister crops that Native Americans grew in New England for centuries before the arrival of Columbus in the New World. Corn is a major cultural icon for most Native American groups. Today in the Northeast the Iroquois or Haudenosaunee still have their native varieties, the most famous of which is the 'Tuscarora', which is also know as 'Iroquois White'. This corn has very large ears and can be 12–14 inches long with fat cobs. It is an impressive sight. Native lines can be obtained in the Seed Saver's Exchange Yearbook and from Rowen White of the Haudenosaunee Seed Collective (see the seed company profiles in appendix B). Lawrence  Davis-Hollander of Eastern Native Seed Conservancy also sells many of these varieties. Iroquois corns have blue, black, red, greenish, pinkish, and yellow kernels and vary from hard flints to flour corns to black sweet corn.

Outside of Iroquois country there are still many heirloom corns in New England, not as varied but definitely worth growing. Recently Tom Stearns of High Mowing Seeds with Anne Miller (SSE member) and other growers rescued a wonderful northern corn called 'Roy's Calais Flint' or just 'Calais Flint'. It is an eight-row corn with large, very luminescent kernels. This corn comes as all red cobs or all yellow cobs; both are beautiful. It is very early and well adapted to the cool Vermont valley where it was grown for generations. In southern New England, there is a long tradition of using 'Rhode Island White' corn for johnnycakes. This corn seems to have its origin from the Pequot Native Americans and is now almost extinct. Harry Records of Exeter, Rhode Island, still maintains this variety; he and the farm staff at Old Sturbridge Village, to the best of my knowledge, are the only people who grow it in any quantity. The nearby Kenyon Gristmill still occasionally

Yellow 'Calais Flint' corn with hand sheller.

grinds batches of this corn for sale and is looking for people to grow it in quantity. Records used to farm with horses but switched to mechanized growing. He harvests with a New Idea ear picker he purchased from Amish country. Jack Lazor of Butterworks Organic Yogurt also uses a New Idea corn picker but thinks that it is too rough on the corn because northeastern flint corns were not bred to withstand mechanical harvest. Records then dries down the ears in his corn crib. The corn crib is a shed in which he has several bins. The sides are somewhat open for air circulation, but he has covered the entire building with chicken wire to keep animals out. He shells and grinds small quantities himself and sells the meal, which has been kept fresh in a freezer. He has also been producing larger amounts of meal for restaurants.

I highly recommend isolation for corn seed saving. Hand pollination is possible, but labor intensive. Corn seed keeps for several years if stored properly. You can rotate your collection, growing a large amount of just one type each year. Corn pollen can be carried great distances, and all types—sweet, pop, dent/field, flint, and pod—can easily cross with each other. Currently several genetically modified field corns are being sold for corn earworm and corn borer resistance as well as herbicide resistance to Roundup. Be sure you are at least one mile away from commercial silage cornfields; even more distance is preferred. Do not allow any corn from nonorganic animal feed, whole corn, or birdseed to go to tassel in your yard or garden as this could also be a source of contamination. If you are an organic grower, GMO contamination could jeopardize your certification.

There are a few other tricks to keeping corn varieties from crossing. My gardens are near my neighbor's sweet-corn patch. Their sweet corn is generally a standard variety, so I grow extra-early dwarf corns that mature far earlier than their sweet corn. The silks are long dead on my corn when their tassels are beginning to shed pollen. In 2004 I was harvesting dry ears from my 'Tom Thumb' popcorn when my neighbor's sweet corn was in tassel. 'Tom Thumb', 'Gaspé Flint', 'Fort Kent Flint', and 'Canadian Orchard Baby' are all good varieties that mature well before most regular season corns. To ensure time separation between the two corns, I plant the early corn about May 15 and the later corn about June 1 so that there is an additional two weeks head start for the early corn. If you are worried about the late-season corn not maturing, I have started early corn seedlings indoors about May 1 and transplanted the same day as I seeded the late corn—May 15. Time isolation is more difficult in northern New England where some years you will not be able to plant before June.

Kernels will often show contamination if pollinated by a different type of corn. The timing trick is not perfect, and there may be slight overlap with the varieties, but if you plant a white corn next to a blue corn, you can often spot any off-types. Blue kernels will be especially obvious in the white cobs. If flint corn shows up in your sweet corn, the kernel when dry will be full and smooth instead of shriveled. Do not plant heirloom sweet corns near field corns because the heirlooms will be unfit for fresh eating. Besides using timing and off-type kernels, you can plant corn closer than a half mile, but I would not ever recommend closer than 100 yards. You could use a wind barrier between varieties such as a heavily forested area, or a field of sorghum. In this case grow dissimilar corns so that you can see the off-types in the first year cobs and rogue them.

Corn is famous for its hybrid vigor, but is prone to inbreeding depression. One hundred plants is minimal to avoid inbreeding depression; 500 plants is better if possible. If you start with a very small quantity of seed, you may suffer from inbreeding depression. Once your population gets larger you may be able to purge the deleterious genes and restore vigor. I have started with one or two ears of some varieties and grown them for several years and gotten decent yields. The more the better, though. Select full ripe, healthy cobs from at least 10 mother plants. Seed is ready when the husks begin to dry down and turn yellow or brown. For sweet corn this is well past the time you eat it fresh. Native American groups traditionally leave corn on the cobs for storage and hang them until needed. I like to shell the corn using cast aluminum hand shellers I purchased from Southern Exposure Seeds. There are also shellers that can be mounted to a bin. Additionally, there are old fashioned hand-cranked machines such as the Cornell No. 2 in which the ear is placed in the top and the naked cob and kernels are ejected out the other side. After you shell the kernels, winnow off the white chaff with your breath or a small fan and store the seeds in glass jars after they are completely dried. When dry, corn seed will shatter when hit with a hammer. If stored properly in glass jars, corn seeds will keep for at least five years. If frozen they will be viable indefinitely. Corn seed hung in a traditional braid is considered good for three years.

Here are some heirloom corns of interest:

> **Northeastern flint varieties:** 'Longfellow Flint', 'Rhode Island White', 'Byron Flint', 'Roy's Calais Flint', 'Charlie Ellis Flint', 'Vermont Flint', 'Fort Kent Flint', 'Gaspé Flint', 'Iroquois White', several Iroquois calico corns, 'Six Nation's Blue', and 'Mohawk Red'.
>
> **Open-pollinated sweet corns:** 'Baxter Sweet Corn', 'Black Pucker' ('Iroquois Black') sweet corn, 'Country Gentlemen', 'Golden Bantam', 'Howling Mob', 'True Platinum', 'Ashworth', 'Luther Hill', 'Hooker's Sweet'.
>
> **Popcorn:** 'Tom Thumb', 'Strawberry'.
>
> **Open-pollinated dents and flour corns:** 'Wapsi Valley', 'Bloody Butcher', 'Mandan Bride', 'Hopi Blue', 'Nokomis Gold', 'Nothstine Flint'.

Small Grains

In the seed-saving world, vegetables are the main focus, and corn is generally the only grain mentioned. In the Northeast we produce very few of our staple grains. Historically the Champlain Valley was a bread-basket exporting wheat. Now almost none is grown there. In this time of global warming and increasing reluctance to burn fossil fuel to trans-

Spelt.

port food, we should try to be more self-sufficient by growing more local grain. I have never grown acreage of any of these crops, but I have been successful with small amounts of rye, Einhorn wheat, triticale, oats, millet, sorghum, and teff. Most of these plants are self-pollinating, so many varieties can be grown together. Matt Rulevich of Wooly Mammoth Farm in Massachusetts has been growing larger plots of 'Sheba' barley and soybeans. Jack Lazor of Butterworks Organic Yogurt is growing spelt and barley near the Canadian border in Vermont. Will Bonsall's Scatterseed Project also preserves many small-grain varieties that he has been growing for years. Dr. Mark Hutton of Maine Extension is involved with growing quinoa in his state. Although it is not a true grain, quinoa and amaranth play a similar dietary role, but with higher protein. Most Northeast farms do not have the acreage or equipment to harvest these crops and compete with large Midwestern farms. We need to look to Europe and Japan where they grow grains in small fields for examples of how to be more self-reliant.

The *Leguminosea* or Legume Family

Beans *Phaseolus vulgaris* and *Phaseolus* sp.
  Life history: Annual
  Pollination: Insects, bees
  Recommended minimum population size: 25
  Isolation: 50 feet. *Note:* Runner and lima beans need much further isolations.
  Seed-borne diseases: *Alternaria alternate*; *Ascochyta* leaf spots; "bald-heads" of seedling; *Cercospora canescens*; anthracnose; yellows/wilt; root rot; angular leaf spot; red nose; *Rhizoctonia* damping-off/stem canker; *Sclerotinia* wilt/stem rot/watery soft rot/white mold; bacterial brown spot; halo blight; common bacterial blight; bean common mosaic virus; cherry leaf roll virus; cucumber mosaic virus; runner bean mosaic virus
  Common pests and diseases: Mexican bean beetle; *Rhizoctonia* crown rot; *Fusarium* rot; *Sclerotinia* wilt/watery soft rot/stem rot/white mold; anthracnose; rust
  Spacing: Row 18"–27"
  Estimated seed yield: Common bean 1,500–2,500 lbs./acre; scarlet runner 890 lbs./acre
  Soil pH: 5.4–6.5

The genus *Phaseolus* has four cultivated species that are grown: *P. vulgaris*, common bean; *P. lunatus*, lima bean; *P. coccineus*, scarlet runner; and *P. acutifolius*, tepary bean.

P. vulgaris is the most commonly known and grown type. Navy, kidney, 'Black Turtle', pinto, 'Maine Yellow Eye', soldier, 'Jacob's Cattle'—all green beans and all wax beans belong here. As with tomatoes, there are so many beans it can be intimidating. Most common beans will grow in our region if they are from an area with similar day length. I once tried to grow wild beans from Mexico that are the likely progenitor to this crop; they grew luxuriant foliage but they never produced a flower. As you move closer to the equator, the day and night lengths are more equal through the year. Tropical plants flower when the days are getting shorter, closer in length to tropical days year-round. For us in New England, day-length-sensitive beans flower in the shorter days of October.

This species is generally self-pollinating, but outcrosses are common in some heirlooms and landraces such as Iroquois varieties, and in 'Vermont Cranberry'. Seed color is the most obvious way to tell off-types; often the beans are different in the first generation. I am unclear on the inheritance of the seed-coat color, but it seems that it shows up immediately when there has been pollination between cultivars. Other characteristics that can be used to detect crossing include plant-growth habit, pod color, and flower color. Off-types should be rogued.

Bean seeds are mature when the pods begin to dry down and turn yellow. If weather is going to be wet, I would advise cutting the plant off at the base or pulling the whole plant and letting it dry down under cover. Bunches of about ten plants can be hung from the rafters of a shed. Bean pods shatter well only when completely dry.

For shelling moderately large batches, spread plants on a tarp in a large barn and step on the pods

Bean plant and open bean pods.

until the beans break free. The seeds, heavier and smaller, tend to fall to the bottom. The bulk material can be picked up off the top and discarded. Screen the beans and remaining debris using two or three screens. Winnow off like-size debris, remove all off-type and moldy beans.

Pole beans are very different from bush beans when grown for seed. In low-humidity regions such as California they can be grown sprawling on the ground and harvested with the same equipment as for bush beans; in the Northeast this is not possible. The climate is too damp, which causes the pods to mold. Grow pole beans on trellises and harvest them as individual beans mature.

Several bean diseases are seed-borne and are visible as discoloration or sunken lesions on the seed. Save only the best-looking seed. Bean seeds can be stored four years under cool, dry conditions.

For harvesting very large amounts—anything over an acre—a combine with a bean head would be excellent. Large equipment of this type is in use in Maine, Québec, and New York where beans and soybeans are grown. Large equipment can shatter beans and crack the cotyledons within the seed. In California large bean-seed operations use a special bean harvester/thresher with parts made of monkey wood, which is especially soft and gentle with the beans. This is a hard piece of equipment to find. In Northeast operations, I have seen a conveyor-belt system where beans are manually inspected for moldy or split seeds.

Peas *Pisum sativum*

Life history: Annual

Pollination: Insect, bee

Recommended minimum population size: 25

Isolation: 50 feet minimal, preferably 100 feet or more

Seed-borne diseases: *Ascochyta* spp.; leaf and pod spot; white mold; anthracnose; powdery mildew; *Fusarium* wilt; foot rot/blight and black spot; downy mildew; foot rot and collar rot; damping-off/stem rot; bacteria blight; purple spot; pea early browning virus; pea enation mosaic virus; pea false leaf roll virus; pea mild mosaic; pea seed-borne mosaic virus; *Ditylenchus dipsaci*

Common pests and diseases: Damping-off complex; mildew; pea weevils

Spacing: Row 16"
Estimated seed yield: 1,000–1,500 lbs./acre
Soil pH: 5.5–6.5

Peas (*Pisum sativum*) are likely one of the oldest domesticated crops. The Fertile Crescent and the Mediterranean are the principal centers of diversity. In addition to green pods, peas can have purple or yellow pods. The plants themselves can vary tremendously from bushy 6- to 12-inch to 6-foot-plus vines. Peas have several different culinary characteristics. Field or soup peas have fibrous pods, and only the seeds are eaten. Snow pea pods, less tough, are eaten young in full shell. Sugar snap types have relatively little fiber in their pods; they fill out and develop a very sweet taste. Almost any pea type can be grown in the Northeast with proper care. We have been successful with 'Golden Sweet', 'Purple Podded', 'Cascadian', 'Sugar Ann', 'Sugar Lode', 'Sugar Snap', and 'Oregon Giant'. My wife recommends 'Cascadian' based on their good flavor.

Peas are self-pollinating. Fifty feet would be adequate isolation, 100 feet even better. I have heard of many pea off-types. Perhaps they cross more than expected, seed batches get mixed, or hidden recessive genes pop up and express themselves once in a while. Often the smaller seed producer is able to spot these; the larger companies do not examine each plant individually. Turtle Tree Seed originally purchased 'Cascadian' snap pea from a commercial source and found many off-types in it. Because their grow-outs are relatively small, they have been able to rogue the incorrect plants and select for correct snap pea type. They believe their line now to be pure.

Peas drying on the vine.

The plants will begin to dry down in mid-July. Great care must be taken that the plant, pods, and seeds do not mold. Turtle Tree Seed grows peas on ½-inch chicken wire that allows the vines to climb but not grow through. This gauge trellis permits easy harvest of the whole plant. They then place the whole plants on screen tables in a greenhouse with the doors open. When the plants and pods are dry, they thresh. Any of the methods described for beans will work: a sack or tarp for small scale, belt or stationary thresher for larger scale. Winnowing works with screens and a fan. Use a clipper mill or fanning mill for slightly larger amounts. A combine is unlikely to work in the Northeast because the plants will not reach the proper state of dryness in the field and will likely mold.

Fava Beans *Vicia faba*

Life history: Annual

Pollination: Insect, bee

Recommended minimum population size: 25

Isolation: Caging; 1 mile

Seed–borne diseases: Leaf and pod spot; chocolate spot; anthracnose; foot rolls/wilt; net blotch; rust; bean yellow mosaic virus; broad bean stain virus; *Echtes ackerbohnnenmosaik* virus; *Ditylenchus dipsaci*/stem eelworm

Common pests and diseases: Chocolate spot; aphids

Spacing: Row 28"; plant 10"

Estimated seed yield: 1,340 lbs./acre (George 1999)

Soil pH: 6.5

Fava beans, like peas, are of Near Eastern origin and have been cultivated for thousands of years. Males of Mediterranean descent sometimes have an inherited condition called *favism*; for people with this disorder, eating fresh beans can cause the destruction of blood cells and possibly be fatal. On the flip side, this genetic condition protects against malaria when combined with eating dried favas, in a wonderful example of coevolution.

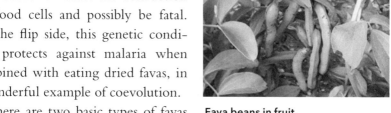

Fava beans in fruit.

There are two basic types of favas based on size: major and minor, the former being large and latter being small. Favas grow quickly, and some can be used as nitrogen-fixing cover crops like other legumes. Most growers in our region do not grow them, though they grow well in the Northeast, and I hope will be grown more widely. In areas with mild winters, they are sown in the fall, but our winters are too cold. The trick is to plant them at about the same time as your peas, as they are a cool-season plant. When the heat of summer comes they begin to wilt, get chocolate spot, and aphids attack them. However, if planted too early, the seed will rot. Will Bonsall, curator of favas for the Seed Saver's Exchange, suggests that anthocyanine, a purple pigment in some darker-seeded types, gives them extra cold-hardiness by preventing the seed from rotting early in the year. These may be good varieties to grow in our climate. I have been successful growing 'Habas Chiquitas', a minor type from Native Seed/SEARCH.

Favas can self- or cross-pollinate. Opinions on isolation distances differ. Some sources say they are mainly self-pollinating with little isolation necessary; others suggest one mile. I recommend a half mile to insure seed purity. Few people grow favas, so probably isolation will not be necessary unless you are growing multiple varieties. Or you could cage them, use row cover, or bag the flowers. The pods dry down to a dark brown or black color when seeds are mature. Like peas, the seeds are ready midsummer and may need to be placed in a shed to dry completely and not mold. Threshing and winnowing are similar to common beans or peas.

Chickpea or Garbanzo *Cicer arietinum*
   Life history: Annual
   Pollination: Insect, bee
   Recommended minimum population size: 25
   Isolation: Caging; ½ mile

The chickpea is another legume that has been cultivated since antiquity. We all know the fat tan-colored type that comes in cans in this country, but there are also black, green, and red varieties. The plants are small-leaved, bushy, upright, and covered with hairs. The pods contain one or two seeds. They can be planted about April 15 in Connecticut.

This plant appears to be mostly self-pollinating, but some studies suggest a fair amount of crossing. I would isolate a half mile between varieties. You are unlikely to have pollen contamination from your neighbors. Plants may also be caged or grown under row cover. I have had success with a green chickpea called 'Green Chana' purchased from J. L. Hudson Seedsman and 'Garbonzo de la norte' from Native seeds/SEARCH. 'Kabouli Black' was a popular variety in the Northeast for a short time.

Will Bonsall grows several chickpea varieties and offers them through the Seed Savers Exchange Yearbook. Seed cleaning is similar to pea or bean, though threshing may be slightly more difficult.

Soybeans *Glycine max*

Life history: Annual

Pollination: Insect, bee

Recommended minimum population size: 25

Isolation: Self-pollinating; 50 feet

Seed-borne diseases: Bacterial blight; brown spot; pod and stem blight; *Sclerotinia* stem rot; downy mildew; soybean cyst nematode; bean pod mottle virus; soybean mosaic virus; tobacco ring spot virus (bud blight)

Common pests and diseases: Japanese beetles

Estimated seed yield: 1,500–2,500 lbs./acre

Soybeans are the protein powerhouse of the plant world. This crop hails from Asia and is known to have been cultivated for several thousand years. There are many types: black, white, and green soybeans. Here in the Northeast few farmers grow the field types of the Midwest and South. In recent years the edamame type has taken off in our region. These are eaten boiled in the pod similar to shell beans. The Japanese eat them while drinking beer, like peanuts in this country. I think they are very tasty with or without beer.

Soybeans are self-pollinating and actually have their pollen shed on their stigmas before the flowers even open. Cross-pollination is unlikely, so many different types may be grown near each other. 'Black Jet' from Johnny's Selected Seeds produced well for me. 'Butterbeans', 'Envy', 'Sayamusume', 'Beer Friend', and 'Shirofumi' are all grown in our area. Soybeans appear to have many fewer disease problems than other beans. There are generally few fungal problems, such as rusts or anthracnose, and the Mexican bean beetles did not seem to like them either. Japanese beetles do attack them, which makes sense since they are from the same region of the world.

When harvested, soybeans should be treated similarly to other legumes. The pods will dry down and turn brown. Soybean pods are much less likely to rot, but take care to keep them dry and free from mold at maturity.

Clean similarly to common beans. Combines are a possibility for large-scale harvesting.

The *Alliaceae* or Onion Family

Onion *Allium cepa*, Leeks *Allium ampeloprasum*, Shallots *Allium cepa*
    Life history: Biennial
    Pollination: Flies, bees
    Recommended minimum population size: 25
    Isolation: Caging; 1 mile or greater
    Seed-borne diseases: Purple blotch; damping-off/gray mold; seedling damping-off/neck rot; collar rot/leaf spot; leaf blotch; smudge; *Fusarium* spp.; downy mildew; black stalk rot/leaf mold; rust; white rot; smut; onion mosaic virus; onion yellow dwarf virus; *Ditylenchus dipsac*; eelworm rot
    Common pests and diseases: *Botrytis* blight and *Pythium*; damping-off; pink root
    Spacing: Row 28"; plant 3"–4"
    Estimated seed yield: Onions: 670 lbs./acre; leeks: 530 lbs./acre (George 1999)
    Soil pH: 6.0–6.8

Onions, leeks, and shallots are biennial and insect-pollinated. Leeks are a separate species and will not cross with onions or shallots. Shallots and onions belong to the same species, and the two can cross. Isolation distances of over a mile between varieties are recommended for commercial seed production. Alternate-day cage methods also work. Onions can sometimes overwinter if mulched properly, but most growers use the seed-root-seed method, placing the bulbs in a root cellar for the winter. Leeks are hardier and can be overwintered, at least in southern New England, but I recommend storing them in a root cellar. 'New York Early', 'Walla Walla', 'Ailsa Craig', and 'Clear Dawn' are open-pollinated onion varieties grown in the Northeast.

Onions are started in February or March of the first year and planted out in the late spring or early summer. The tops will die down in late summer and the onions can be dug. Inspect the bulbs for off-types and set aside the best bulbs for replanting the next year. Onions store nicely in most cool basements, but cannot survive hard freezes. Plant bulbs in the mid to late spring of year two. The tops will send up hardier-looking leaves than the previous year. Flower clusters called umbels will appear in July and be pollinated. Stake umbels to prevent soil-borne disease. When the seed pods mature, clip them and place them under cover to dry as quickly as possible to prevent the seeds from shattering. When they are completely dry, thresh the seeds on a tarp,

Allium cepa

or with the tray and paddle method. Screen and winnow them. Seeds are relatively short-lived in storage, viable only two years.

Garlic *Allium sativum*

Life history: Perennial

Pollination: N/A

Recommended minimum population size: Any

Isolation: None

Seed-borne (bulb) disease: *Ditylenchus dipsaci* stem and bulb nematode; *Sclerotium cepivorum* white rot; *Fusarium oxysporum* basal rot; *Botrytis* neck rot; *Penicillium biferum* penicillium mold; yellow dwarf virus

Common pests and diseases: The above plus pink root and onion maggot; army worms; and wire worm. See *Growing Great Garlic* (Engeland 1991) for more detail.

Garlic *(Allium sativum)* is a wonderful crop that is easy to grow and much in demand. The variability ranges from hard necks with large cloves, soft necks with small cloves, porcelain white to red and brownish. Garlic is fall-planted in late September or October and then mulched. Best time to plant is six weeks before the ground freezes hard so that the garlic makes root growth, but not top growth. The plants come up the next spring and produce scapes, which are odd-looking structures resembling something like birds' heads with a long beak produced on a lengthy stem. The scapes produce bulbils that can be planted to make a garlic bulb in two years. Very few growers use bulbils for planting. Most feel that they need to cut off the scapes, so the plants allocate their resources to produce large bulbs. The bulbs are ready mid-July to early August and are pulled when still partially green (Engeland 1991). You can tell when top-setting garlic is ready to harvest; the cloves begin to separate from the central stalk. Bunch the plants and allow them to cure in a well-ventilated space. Storage fungi will cause problems if garlic is not dried correctly. Store bulbs until the next fall. Garlic has lost the ability to reproduce sexually, so there is no danger of outcrossing, though many types look similar, and good labeling is needed.

The *Solanaceae* or Nightshade Family

Tomatoes *Lycopersicon lycopersicum* (syn. *esculentum*)
  Life history: Annual
  Pollination: Insect, bee
  Isolation: 25 feet
  Recommended minimum population size: 25
  Seed-borne diseases: Early blight; stem canker; leaf mold; *Fusarium* wilt; anthracnose; fruit and stem rot; late blight; buckeye rot; damping-off; *Verticillium* wilt; bacterial speck; bacterial spot; bacterial canker; *Septoria* leaf blight; cucumber mosaic virus; potato spindle virus; tobacco mosaic virus; tomato ring spot virus
  Common pests and diseases: Same as above; flea beetles; tomato hornworm
  Spacing: Row 35"–47"; plant 24"–36"
  Estimated seed yield: Variable; 70 lbs./acre, about ¼–1 oz. per plant, up to 200 lbs./acre
  Soil pH: 6.5

In the height of summer nothing is better than to bite into a ripe, juicy, tasty tomato off the vine. Heirloom tomatoes are astoundingly diverse. They can be yellow, red, pink, bright green, brownish black, and have stripes and streaks of various sorts. Wild tomatoes were originally from Mexico and brought to Europe. From there they came to the United States, many with Italian and German immigrants. Acceptance was slow at first since the tomato belongs to the nightshade family and was thought poisonous. There are tales of people eating a fruit publicly in colonial times to demonstrate that it was edible (Heiser 1969).

Cherry tomatoes.

We have an overwhelming diversity of tomato varieties available to us. Most years maturity is not an issue for most varieties. The biggest challenges are production of the plants, disease resistance, and fruit appearance. Heirloom tomatoes haven't generally been bred for appearance as much as modern varieties, so many of them are lumpy, ruffled, and crack easily when rain comes after a dry spell. Also, some of the beautiful and unique colors may not appeal to consumers who believe only in round red tomatoes. Never fear, some heirlooms fit those consumer ideals and offer the best of both worlds.

The vast array of tomatoes in the Seed Savers Exchange catalog and other seed catalogs makes choosing difficult. If in doubt, ask another local grower. If you have a staple variety, you may want to stick with it and then trial small amounts of interesting tomatoes on the side to see if they fit into your farm's system and markets. One of the best places to get a good look at many tomato varieties is Eastern Native Seed Conservancy's Tomato Fete, held in the Berkshire Mountains of Massachusetts. I have seen over 100 tomato varieties there, all sliced for tasting. Chefs prepare some amazing dishes with them. The tasting is usually held during the first or second week of September. Tickets are limited, but it is worth attending if you want to know what these varieties look and taste like. Amy LeBlanc in Maine sells tomato seedlings for over 130 different heirlooms. In southern New England, Alice's Organic Greenhouse and John Sokoloski in Connecticut offer plants and seeds, respectively. Many greenhouses and garden centers are beginning to carry heirloom tomatoes as seedlings. I have been successful with 'Rose de Berne', 'Brandywine', 'German Johnson', 'Zapotec', 'Black Pear', 'Yellow Perfection', 'Snow White Cherry', 'Ceylon', 'Green

Zebra', 'Striped Cavern', 'Aztec B', 'Extra Dwarf', 'Ethiopian Black', 'Black Plum', 'Persimmon', and 'Peacevine Cherry'.

Tomatoes are considered selfers because their stigmas are usually below the anther cone and are not exposed to pollen from other flowers. Varieties can be grown relatively close together, but I do not recommend it for commercial seed production. I have seen quite a bit of outcrossing in tomatoes when 20 to 30 cultivars are grown within a few feet of each other. Potato-leaf types are especially prone to outcrossing because they have protruding stigmas and often cross to other varieties. Some 'Brandywine' strains are examples of potato-leaf types. Individual varieties should be separated by 50 to 100 feet for a seed crop. Tomatoes do not exhibit inbreeding depression, though I would grow a bare minimum of 6 plants to check for off-types and 25 for selling seed to have a larger genetic sample. Collect fruit from the most robust and disease-free individuals for seed. Collect fruit from as many individuals as possible to maintain genetic diversity. Do not harvest seed from diseased or cracked fruit or fruit on the ground because of risk of soil-borne diseases.

'Brandywine' and 'Rose de Berne' tomatoes.

Tomato seed cleaning is easy. Once the tomatoes are ripe, cut them and squeeze out the gel-covered seed into a yogurt container or a bucket and add enough water to nearly double the volume. If you clean the seed by hand, you will have the rest of the tomato to use in sauces or in salsa. Smaller-fruited types can go in the blender on a slow speed. The Millet Seed Extractor is a small machine for extracting large amounts of tomato seeds. It crushes the tomatoes to get out the seeds.

Ferment the seeds until a light layer of mold forms over the top of the liquid. If the weather is warm you will want to watch carefully because this can happen quickly, in one or two days. Fermentation destroys some diseases and naturally occurring seed-germination inhibitors that prevent the seeds from growing at the incorrect time. If allowed to ferment for too long, tomato seed will start to sprout, or drown. So do not ferment longer than seventy-two hours. Pour off the top part of the fermentation containing fruit debris, mold, and bad seed. Refill the container and decant the liquid off of the seed. Repeat until only clean seed remains. Spread the seed out on a fine mesh screen to be rinsed. I usually leave them on the screen to dry. Stir the seed several times during the first day or two of drying, otherwise the seeds will clump and stick together. Nylon stockings or mesh bags with a small weave work well to dry seeds. Hang the bags to dry or put them in a clothes drier on fluff with no heat until the seed is dry and won't stick together. Do not dry seeds on cloth, paper towels, or paper plates as they will stick to those surfaces and be difficult to remove. Do not place tomato seeds in direct sun or in temperatures above 95°F. Dry down seeds completely before putting them away. I have had tomato seeds keep with very good germination for three to five years in glass jars.

Tomatillos, Cape Gooseberries, Husk Tomatoes, and Ground Cherries *Physalis ixocarapa* and *Physalis* spp.

Allow fruits to ripen and fall off. Tomatillo husks stay green even though the fruit and seeds are ripe. The other species will have dry and tan husks when mature. Follow instructions for tomato seed growing. *Note:* Tomatillos cross-pollinate more often than tomatoes. Use greater isolation distances between different types.

Peppers *Capsicum annuum* and *Capsicum* sp.

Life history: Annual
Pollination: Insects, bees
Recommended minimum population size: 25
Isolation: Caging; 500 feet
Seed-borne diseases: *Alternaria* spp./fruit rot; frog-eye leaf spot; anthracnose; *Diaporthe* fruit rot; bacterial leaf spot; *Fusarium* wilt; leaf mold; *Phoma* rot; *Phytophthora* blight; *Rhizoctonia*; *Sclerotinium* rot; bacterial rot; bacteria leaf spot; alfalfa mosaic virus; tobacco mosaic virus; cucumber mosaic virus
Common pests and diseases: *Phytophthora* blight; bacteria leaf spot; *Sclerotinia* stem rot; viruses
Spacing: Row 18"–35"; plant 12"–24"

Estimated seed yield: 50–100 lbs./acre

Soil pH: 6.0–6.5

Peppers are among the most beautiful vegetables you can grow in your garden. The colors are outstanding: bright orange, red, and yellows balanced by purples, browns, and greens. Peppers were domesticated in Central and South America. A wild chili, the *chiltepin*, range naturally into Texas and Arizona and are preferred in some Mexican dishes. The hot ingredient in hot pepper types is a chemical called capsaicin. Birds, which are the natural dispersers of the seeds, aren't affected by it. Peppers originally developed their spicy kick to keep mammals with grinding teeth away from the chili seeds (Nabhan 1993).

There is controversy over how many species of peppers exist. Generally there are five accepted species, but only two are grown in the Northeast. For more information, see the chapter entitled "How Many Kinds of Peppers Are There?" in *Of Plants and People* by Charles Heiser (1985). In the Northeast we have *Capsicum annuum*, and *Capsicum frutescens*, which may or may not be synonymous with *C. chinense*. Most commonly grown peppers belong to *C. annuum*, including all the sweet peppers, regardless of shape and color, as well as most of the common hot types such as jalapeños and cherry peppers. 'Bolivian Rainbow', habañero, and tabasco types belong to the *frutescens/chinense* group and will not cross with *annuum*. So it is possible to grow a bell type near a habañero type without crossing.

We have been successful growing most types of peppers when we started them in the greenhouse at the appropriate time and treated them properly in the field. The only exception is the wild *chiltepin*, which made no fruit the first year. When we brought them inside as houseplants and transplanted them back out for a second year, they produced abundant fruit. Southwestern peppers have worked for us, but often they produce only a few mature fruits that have enough seed to save for replanting, but not enough for commercial production.

Peppers are insect-pollinated and can both outcross and self-pollinate. Five hundred feet is the recommended isolation distance. Caging or covering with Reemay are quite effective methods to prevent outcrosses. My wife and I have grown a dozen pepper varieties in close proximity covered with Reemay or bridal veil cloth and saw no off-types in subsequent years.

Peppers are ready for seed harvest when they have reached complete ripeness. Peppers change color when ripe. Generally mature peppers are yellow, orange, or red. A few varieties turn black or purple when ripe. Extract seeds by cutting the pepper in half then scraping them off the core. Rinse the seeds in a sieve. Seed cleaning can be challenging with the hot varieties. Wear thick rubber gloves; standard latex gloves such as doctors wear are not an effective barrier. Seeds of small-fruited varieties may be processed in a blender at slow speed with some water. Blend until cores have broken down. The good seeds will sink and the pulp can be poured off; repeat until only clean seeds are left. Use a Millet Seed Extractor for large grow-outs. Rinse seeds on appropriately sized screens and spread out to dry on cookie sheets or screens. Pepper seeds are viable for three years if kept in cool dry conditions.

Eggplants *Solanum melongena*

Life history: Annual

Pollination: Insects, bees

Recommended minimum population size: 6–25

Isolation: 50 feet or caging

Seed-borne diseases: *Alternaria* leaf spot and fruit rot; anthracnose; *Fusarium*; damping-off; *Sclerotinia*; *Verticillium* wilt; eggplant mosaic virus

Common pests and diseases: *Alternaria* leaf spot; flea beetle; *Phytophthora* blight; Colorado potato beetle

Spacing: Row 24"–48"; plant 18"–30"

Estimated seed yield: 50–150 lbs./acre

Soil pH: 6.5

Eggplants (*Solanum melongena*) are a warm-season crop originally grown in Africa and Asia. They made their way into Europe through Spain and Italy. The original types were small and white, hence the name eggplant. Currently in the Western world we have dark purple, bright purple, white, green, and orange in a variety of sizes and shapes.

Eggplants are primarily self-pollinating. Grow a minimum of twelve plants. Fifty to 100 feet is the recommended isolation distance between varieties. Caging or covering with floating row cover also works well if you grow multiple cultivars next to each other. Leave the fruits on the plant well past the time of normal harvest. Purple types turn a brownish color, while lighter types turn yellowish. Grate fruits and place the pulp in a bucket or bowl of water. Squeeze the seeds out; good seed will sink, the pulp and immature seed can be poured off the top. You can also extract seed from the pulp by placing sections of the fruit in a blender, adding water and blending at slow speed. Spread seed out to dry. Seed can last seven years under cool, dry conditions. I have been successful growing 'Diamond', 'Listada di Ghandia', 'Udumalapet', 'Rosa Bianca', and 'Applegreen'. Turkish eggplant represents another species, *Solanum integrifolium*, the tomato-fruited eggplant. Some of the Southeast Asian eggplants belong within this species, such as 'Hmong Red'. It is wonderful for its ornamental value; some of them turn bright red or orange when ripe. I have had good luck with Turkish eggplant; it seems to be doing well in the Northeast since the seed came from a grower in Maine. Seed processing and isolation are similar to the regular eggplant species.

Potatoes *Solanum tuberosum*

> **Life history:** Annual
> **Pollination:** N/A
> **Recommended minimum population size:** Any
> **Isolation:** None
> **Tuber-borne disease:** Many; see *The Potato Field Manual* (Simplot 1985)
> **Common pests and diseases:** Many, see Simplot manual
> **Spacing:** Row 3'; plant 12"–15"
> **Estimated seed yield:** Varies greatly depending on variety
> **Soil pH:** 5.0–7.0

Potatoes originated in the highlands of South America and are probably an ancient hybrid between two wild species of potato. Europeans did not know this crop until after the voyages of Columbus. The Spanish and Portuguese transported them across the Atlantic. Andean potatoes come in every shape, color, and form imaginable: black, purple, red, yellow, lumpy, long, etc. Only a few types were introduced to Europe, and they had to adapt to the day length; most potatoes from South America do not make tubers in our long days of summer. The narrowness of the gene pool introduced was likely the cause of the devastating potato famine. Most potatoes came to this country from Europe and not directly from South America, though there seem to be some exceptions, such as 'Anna Cheeka's Ozette', a fingerling grown by Northwest coastal tribes, and perhaps 'Seneca Cow Horn', which is/was grown by the Iroquois.

Potatoes are propagated asexually, and it may be difficult to sell them legally in some areas. It is easy to carry disease with tubers, though fine seed potatoes can be purchased from local farmers. I

Potato plant with potatoes growing.

highly recommend the *Potato Field Manual* by Simplot for recognizing potato diseases. Commercial seed needs to be certified to be sold; contact your state agricultural extension agency for details. Generally, commercial seed potatoes are grown for only a few generations before returning to a nuclear stock. Viruses tend to cause "running out" in potato stock that reduces the crop yield. This nuclear stock is maintained in a virus-free environment. Seed potatoes can also be cleaned of viruses in tissue culture. Cornell is currently culturing several heirlooms to remove viruses and restore their production vigor.

There is no danger of cross-pollination with potatoes, though there is a good possibility of confusing similar types, so be sure to label your rows well. For seed potatoes, grow potatoes as you would for harvest. Smaller tubers about the size of a chicken egg are good to save; keep them in a root cellar and try to keep them from sprouting. Pre-exposure to light, called "greening" or "chitting," is a common practice in Europe. It allows the potato to get a head start before it is placed in the soil. I have found late planting in mid-June to early July in Connecticut avoids Colorado potato beetle, and the late harvest ensures the storage life of the potato to the next season. Many growers purposefully do not wash their potatoes until they are ready for use, especially white-skinned types that can develop silver scurf after they are washed.

Dill.

The *Umbelliferae* or Carrot Family

This aromatic family includes dill, cilantro, celery, parsnips, and carrot. They all have umbrella-shaped groups of flowers called umbels. Most of these plants are biennial and generally cross-pollinating. Use alternate-day caging techniques, or bag individual umbels.

The same general hand-pollination technique can be used for all umbels. First bag the flowers before they open. Lift the bag and brush the palm of your hand over the flowers of one plant, covering your hand with pollen. Then carry the pollen to another plant of the same variety and brush the palm of your hand on this plant. In the process you will get pollen from this plant on your hand as well as transfer pollen from the first plant to the second. Take this pollen mixture and rub it on the other plants of this variety, and be sure to go back and place the mixed pollen on the plant you started with. After you apply the pollen, replace the bags as soon as possible.

Root Umbellifers: Carrots *Daucus carota* and Parsnips *Pastinaca sativa*

Life history: Biennial
Pollination: Insects, especially bees, flies
Recommended minimum population size: 25
Isolation: Caging, ½ mile or greater
Seed-borne diseases (carrot): *Alternaria* leaf blight; black root rot; bacterial leaf blight; *Cercospora* leaf blight; powdery mildew; *Phoma* root rot; *Xanthomonas* leaf blight; carrot motley dwarf viruses
Common pests and diseases (carrot): *Xanthomonas* leaf blight; *Sclerotinia* blight; tarnished plant bug

Seed-borne disease (parsnip): *Alternaria* leaf blight; black mold; powdery mildew; canker; *Phomopsis diachenii*; *Pseudomonas viridiflava*; strawberry latent ring spot virus

Spacing: Carrot, row 30"–36"; plant 8"–12"; parsnip, row 40"; plant 12"–24"

Estimated seed yield: Carrot 530 lbs./acre; parsnip 890 lbs./acre (George 1999)

Soil pH: 6.5

Carrots are the most widely known and grown member of this family in the United States. In addition to orange roots, there are varieties of white, yellow, maroon, and purple-black colored carrots. Carrots are biennial and can be grown seed-seed or seed-root-seed. Most carrots are hardy in our region with protective mulch and can be left in the ground to overwinter. I like to look at the roots to check for off-types. Select the carrots and rogue off-types, then store the good roots in a root cellar to be planted out the next year. By looking at the roots you can rogue off-types before they flower, and pollinate true-to-type individuals, selected for paternal (pollen) parents as well as maternal (seed) parents.

Carrot varieties are pollinated by syrphid flies, houseflies, bees, and butterflies; leave a half mile between varieties. Use bagging or caging to isolate different varieties grown in close proximity. Place a bag over the flowers, then hand pollinate them. Use an alternate-day caging method to allow cross-pollination within the cultivar. BEWARE! Queen Anne's lace, a very common weed, is the same species as carrot but without the nice large orange roots. It can easily cross into your carrots and ruin the next generation. If pollinated by Queen Anne's lace, the carrot roots will be white, fibrous, and small. Isolation of a half mile or more is required between Queen Anne's lace and carrot. Harvest seed when the umbels brown. A 1/22-inch screen is recommended to clean seed away from larger chaff. Winnow with care (carrot seed is light) using a length separator, followed by an air column. Enjoy the wonderful aroma of carrot seeds when you clean them!

Parsnip (*Pastinaca sativa*) is a sweet root crop that should be grown more widely. They are definitely a treat in midwinter. Most techniques for growing, pollination, and seed cleaning are the same as for carrot. Like carrot, parsnip has a weedy wild relative (don't we all have relatives like that!). Wild parsnip seems to be common in cooler areas of the Northeast, especially if the soil is influenced by limestone bedrock. The Champlain Valley of Vermont, the Berkshires, and Litchfield County in Connecticut have abundant wild parsnip populations. One mile is recommended between parsnip varieties and also between cultivated and wild parsnip. Juice from parsnip stems can cause chemical burns; wear gloves and long sleeves when working with the plants. Parsnip seeds are notorious for their short shelf-life; keep them dry and cool; even then they may last only two years. Freezing thoroughly dried seed in glass jars, seal foil, or plastic envelopes will greatly extend the storage life of parsnip seeds.

Cilantro *Coriandrum sativum*, Dill *Anethum graveolens*, and Parsley *Petroselinum crispum*

Seed-borne disease (parsley): *Alternaria* spp.; leaf and stem rot; powdery mildew; brown root rot; *Phoma* sp.; root and basal stem rot; leaf spot; parsley latent virus; strawberry latent ring spot virus

Spacing (parsley): Row 36"; plant 10"

Estimated seed yield (parsley): 712 lbs./acre (George 1999)

Estimated seed yield (dill and cilantro): 1,000 lbs./acre

These three herbs are also umbellifers, and similar pollination and seed cleaning techniques apply. Cilantro grows well during cool seasons and if spring-planted should bolt and flower in one year. Dill is a warm-season annual. Parsley is a biennial that can be overwintered in all but the coldest areas of the Northeast; a good mulch protection is recommended. They are all outcrossers, and the seeds turn brown in the umbel when they are mature.

Appendices

A. Seed-Cleaning Devices for Cleaning Large Lots of Seed

The following tools may be of use to growers who are planning to grow very large quantities of seeds. All are currently in use in the Northeast.

Millet Seed Extractor—great for crushing the fruit of peppers and tomatoes to extract the seeds.

Liberty 5000 Wet Vegetable Seed Extractor—used to extract melon, watermelon, cucumber, and squash seeds. It crushes the fruit, then spins the pulp around on a tumbler screen and the seeds drop out onto a platform below.

Length separators—cylinders or discs with divots of a specific size that pick up seeds that fit within them and carry them to a bin. These can be used to remove broken seeds or chaff.

Bean tables—conveyor belts that carry beans along so they can be inspected for "splits" or mold.

Clipper mill—a machine with a series of screens to separate out seed from chaff, but also with a blowing fan, thus an improvement over the old fanning mill.

Velvet rollers—these separate out round, full seed from broken seeds. The smooth full seed slides down the middle of the two rollers while the broken seed catches on the cloth and is thrown outward.

Air columns and blowers—plastic cylinders that separate chaff out by blowing it up and into side pockets. Conversely, if seed is full of small pebbles, the air pressure could be turned up to blow the seed up into the pockets while the pebbles remain in the plastic cylinder.

Spiral—a metal spiral, shaped like an auger. Seeds are placed in the top and high-germination round seeds accelerate and spill off the side, cracked or withered low-germination seeds fall out the bottom.

Combine—a machine that harvests, threshes, and winnows seeds. Combines are manufactured by most companies that make farm equipment. They have different types of heads mounted on the front for harvesting different crops.

Fanning mill—a small, usually hand-driven machine; a fan in the back creates a breeze to blow away chaff while seed passes through various vibrating screens. Clean seed falls to the bottom. There are mechanized versions of the old style. The clipper mills are the most recent and currently available type. Fanning mills are common in antique stores or at barn sales.

Screens—can be fancy or basic. You can buy hardware cloth or window screen and attach it to a frame. Old hand screens can occasionally be found in every size (gauge) imaginable. Clipper mill screens can be put into homemade frames for hand-screening use.

B. Seed Companies and Organizations Devoted to Heirloom Crop Preservation in the Northeast

Turtle Tree Seed

The Northeast is fortunate to have Turtle Tree Seed, the only fully biodynamic vegetable seed company in the United States. Run by Nathan, Beth, and Yarrow Corymb, it is located in Camp Hill, a community dedicated to living and working with special-needs people in Copake, New York. The Corymbs were sponsored by the Biodynamic Association to spend over a year studying in Switzerland to learn biodynamic seed-growing techniques. Through their European connection, they have brought back unique varieties and, in many cases, are the only company offering them for sale in this country. The Corymbs estimate they offer thirty such unique varieties. While visiting them I sampled one from Switzerland, the 'Schweizer Riesen' or 'Swiss Giant' snow pea. It was very large and tasty.

I was impressed by this small company's dedication and thoroughness. They maintain varieties with extreme attention to detail. They constantly select for quality and consistency among the plants within each variety they offer and strive to offer superior versions of each in their catalog. They are also testing strains of different varieties. They have found great differences among different strains of 'French Breakfast' radish and

'Scarlet Nantes' carrot. Beth and Nathan claim not to be plant breeders, but I believe that they do cross that line, in the same way that many other seed savers do. Their area has a high incidence of cucurbit bacterial wilt, and they have been selecting their 'Dark Green' zucchini for resistance to this disease by roguing any plants that begin to wilt. This is an example of breeding for horizontal resistance. There is no known single-gene vertical resistance for bacterial wilt.

The Corymbs have all types of small-scale hand tools for cleaning. Their shop houses standard cleaning equipment such as screens, air-separation columns, clipper mills, and amazing numbers of hand screens. Additionally, they have unique, low-tech but ingenious tools like a wooden tray with an uneven base brought from Switzerland. The seeds are set in the deeper part of the tray, then a fanning motion creates an air current that winnows off the chaff. It is a simple and effective tool for cleaning chamomile seed. Other unique and simple tools are ceramic trays with 25-, 50-, or 100-count indentations for the "differently-abled" residents of Camp Hill village so that they can participate in packet filling for Turtle Tree Seed. Conversely, their seed workshop has extremely high-tech aspects as well, such as climate-controlled chambers for germination tests, a cooled, dessicant-dehumidified trailer for seed storage, a custom-built stationary thresher from Europe, and length separators.

High Mowing Seeds

I remember receiving a small photocopied handwritten catalog from Tom Stearns in 1996. His company was then called the Good Seed. A few years later he settled in northern Vermont, changed the company name to High Mowing Seeds, and began collecting varieties grown by local seed savers, among them 'Calais Flint' corn. High Mowing is 100 percent organic, offers some biodynamic seeds and a large proportion of locally-grown seed. Over the past few years it has gone from a small one-man operation to a major player in the seed business in the Northeast, right on the cutting edge of research and production. Stearns is always experimenting with new varieties to test for flavor, production, and disease resistance. I went with him to his research plots and saw replicated rows of romaine lettuce, Swiss chard, and spinach. He and his staff know only the numbers, so names of varieties do not bias the results. In his trials, he started with thirty different spinach varieties, narrowed to the best eight, which he trialed again the next year. He will offer the best ones in his catalog.

Stearns has built a small lab in his trailer, which was his bedroom at one time, and has hired a plant pathologist to research plant disease and check that the seed is of the highest quality and disease-free. Tom is very knowledgeable about seed-borne diseases.

High Mowing is also doing research on optimal planting time and spacing needs of tatsoi and mizuna. Tom says, "I love growing brassicas," and his seed-cleaning machinery shows it. He has a Massey-Ferguson combine with which he harvests mustard-green seeds. Tom has become a seed-cleaning equipment guru. Bean tables, gravity tables, a seed spiral, and a collection of fanning mills dot the barn he rents as a workshop. He has begun to find and sell equipment to his growers so that they can deliver him a better product. High Mowing has entered a partnership with the Liberty Tool Company of Marshfield, Vermont, to manufacture a wet seed extractor for melon, cucumber, and squash that can process two to three acres of squash in a day. This extractor will likely be on the market in the next few years.

Haudenosaunee Native Seed Collective

The seed collective is a group of indigenous and nonindigenous growers committed to restoring Northeast native heirloom crops. Their collection of rare locally adapted landraces includes purple corn, zebra striped beans, giant orange hubbards, and other wonderful foods and flowers. Rowen White, coordinator and curator for the project, is working arduously to restore small lots of rare native seed stocks into healthy populations. Many HNSC growers are trying to provide enough stock seed to gift back to the Haudenosaunee (Iroquois) people for use in community gardens, large-scale growouts, and more. Honoring the cultural and historical connection to these seeds, Rowen conducts seminars and workshops from elementary school to college level to share the rich heritage of her people in ways that inspire each of us to be stewards of the earth (White 2002).

Will Bonsall and the Scatterseed Project

Will is famous for standing up at meetings and saying, "I make the sun rise in the morning and set at night; other than that I'm just a regular guy." Though he is joking, he does have superhuman seed-saving abilities. Will is curator of potatoes, Jerusalem artichokes, and fava beans for the Seed Savers Exchange. He also has extensive collections of small grains, peas, chickpeas, and scion wood for fruit trees. When I visited his farm, I saw sixty pole bean varieties grown around sunflower stalks, hardy kiwis, and I tasted a wonderful currant variety named 'Cherry'. He conserves close to 3,000 different cultivars. He hopes that his operation is an ark and that the "forty days and forty nights" will someday be over, meaning that the social and economical climate will change and many of his varieties will go back to the regions they came from and be grown by the culture they developed with. Will is in the unique position of seeing, growing, and eating all these varieties. When he identifies a good variety, the seed world listens. Currently Cornell University and NOFA–NY have been inquiring about different potatoes he grows; a rare heirloom may have commercial possibilities.

Besides his conservation efforts, he grows much of his own food "veganically" without using animal manures, enriching the soil instead with leaf compost and cover crops. The Scatterseed project has contributed to the diversity of seeds offered in northeastern seed catalogs and is incredibly important for its role conserving varieties for the future.

Jack and Anne Lazor, Butterworks Organic Yogurt

Butterworks Organic is known mostly for its great yogurt and cream. Jack and Anne Lazor have a wonderfully friendly herd of Jersey cows. In the summer the Lazors pasture the cows using rotational grazing techniques. They supplement with some grain. For grain, Jack has experimented with 'Wapsi Valley' corn, 'Calais Flint' from High Mowing Seeds, and 'Six Nations Blue' corn from the Haudenosaunee Native Seed Collective. Since their operation is so close to the Canadian border, their eyes and minds are pointed north. Pioneer Hi-Bred in the United States does not have a variety suitable for his climate. Jack does grow '3921' from Pioneer Canada but says that "process is my religion" and he would like to be more self-sufficient by growing an open-pollinated corn. The OP corns, though, so far lack the stand-ability of hybrid corn, meaning they often fall over or lodge. He says his New Idea corn picker can also damage the crop.

In addition to experimenting with open-pollinated corns, he grows dry beans and small grains. Lege is a Canadian variety of barley he grew, and he learned from experience recently that a germination test should be done even when you are planting your own seed. This variety, though usually reliable, germinated at 60 percent. He also grows spelt—from which he will mill and sell flour and save some of the leftover grain for seed—35 acres of rye, some Riggadon and Vermont Oats, plus A.C. Barrie bread wheat. He saves seed from all of these. He harvests his small grains with a John Deere combine with a small grain head. Jack also produces dry beans such as 'Black Turtle' and 'Jacob's Cattle' for sale and seed. Much of his equipment is on a larger scale than that of most seed savers, but it works on the same principles. Butterworks is on the right path to growing table grains here and finding OP corn adapted to our climate.

C. R. Lawn and Fedco Seeds

Fedco Seeds grew out of the Maine Federation of Coops in 1978. C. R. Lawn got the idea for a seed cooperative after doing some math and realizing that he could save gardeners and farmers money by buying large volumes of seed and breaking them down. The seed cooperative's popularity spread like wildfire, and within a few years Fedco was filling orders throughout the Northeast. The plain-enveloped and newsprint catalog reflect the company's commitment to saving money for growers. In fact, Fedco did not have a catalog until 1989; that year they set a goal "to introduce a few heirlooms not widely available." Through the years, Fedco has branched out into organic grower supplies, books, trees, and Moose Tuber seed potatoes. Fedco also began producing and purchasing local seeds. Nikos Kavanya coordinates the trial gardeners and seed growers and offers unique but profitable varieties. In recent years Fedco has kept its finger on the pulse of the political world as well, steering clear of GMOs and pointing out that, like the major seed companies, the organic cooperative food suppliers have been going through increasing consolidation.

C. R. Lawn received a law degree from Yale University before moving to Maine and has been working in cooperative ventures ever since. He has a wealth of knowledge about varieties and gardening, plus he just enjoys eating good food. At his farm in Massachusetts he is trialing lettuce, tomatoes, cucumbers, peas, winter and summer squashes, and sweet corn for Fedco. He also produces seed of an edamame soybean and 'True Red Cranberry', a New England heirloom pole bean. I have been fortunate to work with him and his partner Elisheva Kaufman on the Restoring Our Seed project—dedicated, like this handbook, to training farmers to save seed, select it for crop improvement, and breed it for horizontal resistance to disease.

C. Hot-Water Treatment of Vegetable Seeds

The following information is adapted from the University of Connecticut IPM Web site; the Ohio State University Extension Web site; the Public Seed Initiative handbook; and the "New York State Vegetable Conference and Berry Growers Meeting Proceeding" (pp. 193–94).

Several seed-borne diseases of vegetable crops can be reduced or eliminated by hot-water seed treatment. Even if seed-borne disease is just suspected, it may be worthwhile to treat the stock seeds to avoid infection of an entire seed crop. In addition to hot-water treatment, seeds can be surface sterilized with a dilute bleach solution (check with certifier). Hot-water treatment is generally more effective than bleach because the high temperatures can penetrate into the seed and kill any diseases within. Each crop has a specific temperature limit, above which the treatment begins to kill the seed, severely reducing the germination rates. Older seeds are especially prone to heat damage. It is wise to heat-treat a small test batch of seeds followed by a germination test to ensure that you will not damage the entire seed lot.

Seeds are placed in a cloth bag or cheesecloth with a weight such as a sinker or bolt to fully immerse the seeds in water. Use an accurate thermometer. The seeds are preheated in 100°F water for ten minutes. The seeds are then transferred to the appropriate temperature water—keep the temperature as constant as possible; see chart on the following page. Cool water should be kept nearby to add if overheating occurs. The seeds are kept in the water for the recommended time and then removed, cooled under tap water, and spread out to dry at about 70°F–75°F.

Hot-Water Treatment of Vegetable Seeds		
Crop	**Temperature °F**	**Time in Minutes**
Broccoli	122	20
Brussels sprout	122	25
Cabbage	122	25
Cauliflower	122	20
Carrot	122	20
Celery	118	30
Collard	122	20
Cucumber	122	20
Eggplant	122	25
Kale	122	20
Lettuce	118	30
Mustards and Radish	122	15
Pepper or	122 125	25 30
Spinach	122	25
Tomato or	122 125	25 20

D. Screen Sizes for Seed Cleaning

This chart from Clipper Office Tester is meant only as a guide to help seed producers select screens. Individual cultivars will vary in seed size. The size range may also help those who make their own screens. The upper screens separate larger debris from the seeds, the lower screen catches the seed and allows smaller debris to fall out the bottom. Since the measurements are sometimes difficult to interpret, here are some examples to help you read the chart: 1/20 is a screen with round holes that are 1/20 of an inch in diameter, 11/64 is a screen also with round holes but with a diameter of 11/64 of an inch, 11/ 3/4 or 11/64 × ¾ is a slot or oval-shaped hole that in one dimension is ¾ inch and in the other is 11/64 inch. Additionally there are wire-mesh screens that will have a number such as 6 × 20, meaning 6 wires per square inch in one direction and 20 per square inch in the other direction. A "shoe" is the location where the screens are mounted in the machine.

(*Note*: The Clipper Office Tester company has a more extensive chart than the one here for seeds not covered in this manual).

Screen Sizes for Seed Cleaning

Crop Screen	Bottom Screen Upper Shoe	Top Screen Upper Shoe	Top Screen Lower Shoe	Bottom Lower Shoe
Cranberry beans	32	14/64 × ¾	30	16/64 × ¾
Great Northern beans	26	10/64 × ¾	24	11/64 × ¾
Red kidney beans	30	13/64 × ¾	28	14/64 × ¾
Navy or pea beans	22	10/64 × ¾	20	11/64 × ¾
Pinto beans	26	9/64 × ¾	24	10/64 × ¾
Black Turtle	22	7 × ¾	20	8 × ¾, 9 × ¾
Yellow Eye beans	24	11/64 × ¾	22	12/64 × ¾
Austrian Winter peas	18	9/64 × ¾	17	10/64 × ¾
Canada field peas	20	8/64 × ¾	18	9/64 × ¾
Chickpeas, garbanzos	30	11/64 × ¾	26	12/64 × ¾
Soybean, small 2,800–3,200 seeds/lb.	20	9/64 × ¾	18	10/64 × ¾
Soy 2,500–2,800/lb.	22	10/64 × ¾	20	11/64 × ¾
Soy 2,200–2,500/lb.	24	11/64 × ¾	22	12/64 × ¾
Soy, very large 2,000–2,200/lb.	26, 28	12/64 × ¾	26	13/64 × ¾
Beet	22		20, 19, or none	8,9
Broccoli			7	1/16, 20 × 20
Cabbage	8		6, 7, 3/64 × 5/16	1/16, 1/17,
Cucumber	18		16, 17	8, 9
Carrot	7		6, 5½	6 × 26, 6 × 30
Cauliflower			6	1/17, 1/18
Dill			10	1/18
Lettuce	6 × 18	19, 4 × 20	4 × 18, 4 × 22, 24 × 24	20 × 20 1/8" screen
Muskmelon (Cantaloupe)			16	9
Onion	10		8, 9, 10	1/14, 1/15
Pepper			16, 18	6 × ¾
Radish	10		9	1/13, 1/14
Rutabaga			6, 1/18	1/14, 1/16,
Squash			32, 34, 36	23, 24
Squash, butternut			28	16, 18
Tomato			10, 11	1/12, 6
Turnip	1/14		1/16	1/20
Watermelon			24	16
Watermelon, Sugar Baby			20, 12 × ¾	12
Watermelon, Garrison			26, 13 × ¾	19

Resources

Seed-Cleaning Equipment
(from Public Seed Initiative handbook)

Air Column
Hoffman Manufacturing
P.O. Box 547
Albany, OR 97321
(541) 926-2920
www.hoffmanmfg.com

Belt Thresher
Almaco
99 M Avenue/Box 296
Nevada, IA 50201
(515) 382-3506
www.almaco.com

Clipper (Clipper Office Tester)
Ferrell-Ross
785 South Decker Drive
Bluffton, IN 46714
(260) 824-3400

Velvet Rollers
W. A. Rice Seed Company
1108 W. Carpenter Street
Jerseyville, IL 62052
(618) 498-5538
www.wariceseed.com/products.
html

Wet Vegetable Seed Separator
Millet's
P.O. Box 271
2330 Los Robles Road
Meadow Vista, CA 95722
(530) 878-2325

Other Suppliers
Anchor Paper Company (for germination paper)
480 Broadway
St. Paul, MN 55101
(800) 652-9755
www.seedpaper.com

Commodity Traders International
101 East Main St.
P.O. Box 6
Trilla, IL 62469
(877) 735-4655

Glen Mills, Inc.
395 Allwood Rd.
Clifton, NJ 07012
(973) 777-0777
www.glenmills.com

**Idaho-Oregon Mill Supply
Company**
1910 7th Ave. North
Payette, ID 83661
(208) 642-1101
(800) 426-1694
www.iomsc.com/agricultural.htm

**Round Mesh/Metal for Screens
McNichols Company**
45 Power Rd.
Westford, MA 01886-4111
and
2 Home News Row
New Brunswick, NJ 08901-3602
(800) 367-5819
www.mcnichols.com

**Length Separators
Morgan Manufacturing Inc.**
RR 3 Box 178
Owatonna, MN 55060

Oliver Manufacturing
P.O. Box 512
Rocky Ford, CO 81067
(719) 254-7814
(888) 254-7813

Seedburo Equipment Company
1022 W. Jackson Blvd.
Chicago, IL 60607
(800) 284-5779
www.seedburo.com/

**Heirloom Seeds and Information
Alice's Organic Greenhouse**
(tomato seedlings)
Alice Rubin
167 Bender Rd.
Lebanon, CT 06249-1115
(860) 423-4906

Baker Creek Heirloom Seed Co.
2278 Baker Creek Rd.
Mansfield, MO 65704
www.rareseeds.com

Bountiful Gardens
18001 Shafer Ranch Rd.
Willits, CA 95490
www.bountifulgardens.org

Comstock, Ferre & Company
263 Main St.
Wethersfield, CT 06109
www.comstockferre.com

**Eastern Native Seed
Conservancy**
P.O. Box 451
Great Barrington, MA 01230
www.enscseeds.org

Even'Star Organic Farm
48322 Far Cry Rd.
Lexington Park, MD 20653

Fedco Seeds
P.O. Box 520
Waterville, ME 04903-0520
www.fedcoseeds.com

Haudenosaunee Seed Collective
Rowen White
rowenwhite@yahoo.com

High Mowing Seeds
813 Brook Road
Wolcott, VT 05680
www.highmowingseeds.com

J. L. Hudson, Seedsman
Star Route 2 Box 337
La Honda, CA 94020

Irish Eyes & Garden City Seeds
P.O. Box 307
Thorp, WA 98946
www.irish-eyes.com

Johnny's Selected Seeds
955 Benton Ave.
Winslow, ME 04901-2601
www.johnnyseeds.com

Kokopelli Seed Foundation
Dominique Guillet
335 Pony Trail
Mount Shasta, CA 96067
www.kokopelli-seed-foundation.com

Maine Seed Saving Network
P.O. Box 126
Penobscot, ME 04476

Native/Seeds SEARCH
526 N. Fourth Ave.
Tucson, AZ 85705
www.nativeseeds.org

Old Sturbridge Village
One Old Sturbridge Village Road
Sturbridge, MA 01566
www.osvgifts.org

Organic Seed Alliance
P.O. Box 772
Port Townsend, WA 98368
www.seedalliance.org

Public Seed Initiative
www.plbr.cornell.edu/psi/index.html

Peace Seeds
2385 Southeast Thompson Street
Corvallis, OR 97333

Peaceful Valley Farm Supply, Inc.
P.O. Box 2209
Grass Valley CA 95945
www.groworganic.com

Peters Seeds and Research
P.O. Box 1472
Myrtle Creek, OR 97457-0137

Pinetree Garden Seeds
P.O. Box 300
New Glocester, ME 04260
www.superseeds.com

Ronnigers Seed Potatoes
Star Route
Moyie Springs, ID 83845
www.ronnigers.com

Restoring Our Seed
www.growseed.org

Sand Hill Preservation
1878 230th Street
Calamus, IA 52729
www.sandhillpreservation.com

The Scatterseed Project
Will Bonsall
39 Bailey Rd.
Farmington, ME 04938-4321

Seeds of Change
P.O. Box 15700
Sante Fe, NM 87592
www.seedsofchange.com

Seed Savers Exchange
3076 North Winn Rd.
Decorah, IA 52101
www.seedsavers.org

Southern Exposure Seeds Exchange
P.O. Box 460
Mineral, VA 23117
www.southernexposure.com

Territorial Seeds
P.O. Box 158
Cottage Grove, OR 97424-0061
www.territorialseeds.com

Turtle Tree Seed
Camphill Village
Copake, NY 12516

United States Department of Agriculture ARS-GRIN
www.ars-grin.gov/npgs/

Whitehill Farm (tomato seedlings)
Amy LeBlanc
P.O. Box 273
East Wilton, ME 04234
(207) 778-2685
amy@whitehillfarm.com

Wild Garden Seeds
Frank Morton
P.O. Box 1509
Philomath, OR 97370
(541) 929-4068
www.wildgardenseeds.com

Glossary

Accession—an individual sample of seeds, usually used in reference to seed in a seed bank.

Allele—an alternate version of a gene, e.g., an allele for blue eyes, or an allele for brown eyes.

Annual—plant that completes its life cycle in one year, e.g., most garden crops: corn, squash, beans, lettuce, tomatoes.

Anther—top portion of the stamen, the male reproductive structure that produces pollen.

Biennial—plant that grows during its first year and reproduces the second, e.g., carrots, parsnips, beets, cabbage, cauliflower, Brussels sprouts.

Bolt—upright growth and flowering of a plant that has a basal or low-growing vegetative form. To go to seed.

Chaff—the unwanted debris, e.g., pods, flower parts, etc., left over after threshing seeds.

Clonal plants—plants that reproduce asexually from cuttings or divisions, e.g., potatoes and sweet potatoes.

Crossing or **cross-pollinating,** a.k.a. "outcrossing"—transfer of pollen from the anther of one plant to the stigma of a flower on another plant.

Cultivar—shorthand for cultivated variety.

De-hybridization—the process whereby a plant breeder selects individual plants from succeeding generations of a hybrid variety (F2, F3, F4, etc.) until a true-breeding open-pollinated variety is developed.

Disease screening—plant breeding technique that tests for disease resistance by inoculation of plants with spores, viruses, or bacteria. Varieties, strains, or lines that consistently remain healthy after inoculation are considered resistant or tolerant to the disease.

Dominant allele—the form of a gene that is expressed in a heterozygous individual, this allele masks the effects of the recessive allele.

F1, F2, etc.—comes from the Latin *filius* (son), the generations that come after a controlled cross-pollination. Hybrid garden varieties are F1.

Fertilization—union of sperm and egg that forms a zygote, leading to the eventual formation of a plant embryo contained in a seed.

Gene—a unit of inheritance. A portion of genetic material that controls a specific trait.

Gene pool—the collective genetic information contained within a population of sexually reproducing organisms, especially a diverse group of plants that express several traits and have not been fully standardized and stabilized through selection.

Genetic erosion—the loss of genetic traits and resources.

Genetic richness or **diversity**—the complete range of genetic traits within a species.

Genetic segregation—genetic recombination of traits that occurs after an F1 hybrid is allowed to go to seed.

Genotype—genetic makeup of an individual, the totality of genes, including recessive traits that are not expressed.

Germination—start of plant growth within the seed, usually confirmed when the first root, or radical, emerges from the seed coat.

Guild/pollination guild—a group of plant species that flower at the same time and have a similar set of pollinators.

Heirloom crop—an open-pollinated crop variety that has been cultivated for generations by a family, or generally any open-pollinated variety that predates 1945.

Heterozygous—having two different alleles for a trait at the same locus.

Homozygous—having two copies of the same allele for a trait.

Horizontal resistance—a type of resistance to disease in which many genes, controlling multiple mechanisms and traits, decrease infection by the pathogen. This resistance reduces the rate of infection and is generally durable, effective for the indefinite future.

Hybrid variety—a variety created by the crossing of two different inbred lines. Seeds taken from F1 hybrids are not sterile but will not breed true.

Imperfect flower—a flower that is either entirely male or entirely female.

Inbreeding—mating among related individuals.

Inbreeding depression—loss of vigor in a population due to related individuals in a small population pollinating each other and causing detrimental traits to be expressed.

Isolation—separation of a variety or group of plants from another so that mating among or between groups is prevented to insure genetic purity.

Landrace—a population of crop plants developed by farmers that is heterogeneous (several heights, colors, different disease resistance, etc.) but adapted to local environmental and socioeconomic conditions. Most third world farmers and indigenous peoples traditionally use(d) landraces.

Line—a distinct lineage of plants that has been selected for certain characteristics, often by inbreeding or selfing one or a few individuals.

Open-pollinated—any population of crop plant that breeds true when randomly mated within its own variety. Like begets like, though always with some minor variation.

Peduncle—the portion of a plant stem that attaches to a flower or fruit. Peduncle traits are used to differentiate squash species.

Perfect flower—a flower that contains both male and female parts.

Phenotype—the physical manifestation of a plant's genetic material interacting with the environment; the outward appearance of a plant or other organism.

Pistil—female part of a flower made up of the ovary, style, and stigma. When mature, the ovary becomes a fruit.

Pollen—dustlike material in the anther of a flower that contains the male gamete, which fertilizes ovules within the pistil.

Recessive allele—the allele masked by the dominant allele. The effects of this form of a gene are not expressed unless an individual has two copies of the same allele.

Roguing—to remove plants that do not fit the description of the variety; to remove plants that exhibit unwanted traits.

Screening the chaff—use of several different-sized (gauge) screens to separate debris from seeds. Debris of similar size to the seed is removed by winnowing.

Segregation—production of offspring with different phenotypic characteristics from an individual line or variety due to reshuffling of alleles.

Self-incompatible—a characteristic of a plant that prevents self-fertilization. Describes a plant that has the ability to recognize its own pollen and prevent it from germinating or allowing the pollen tubes to reach the ovules.

Selfing—self-pollination, where pollen produced by a plant lands on the stigma of a flower of the same individual.

Self-pollinating—describes a plant that the majority of the time reproduces by use of its own pollen, a.k.a. "selfer."

Stamen—male or pollen-producing part of the flower, composed of the anther and filament.

Stigma—very top of the female portion of the flower, the part of the flower that receives the pollen.

Strain—subvariety or unique variant of an open-pollinated variety.

Threshing—breaking apart of flower parts, fruit pieces, and other debris from the seed.

Trait—a quality such as color, size, growth habit, pest resistance, and flavor expressed by a plant.

True to type—describes a plant that exhibits the expected qualities and traits of a variety.

Variety—a population of genetically related crop plants that is uniform for a set of traits or characteristics.

Vernalization—cold treatment to induce flowering in a biennial.

Vertical resistance—resistance that is the result of a single gene or a very few genes. The gene produces large effects (i.e., the plant is either resistant or is not resistant), and the effects often are not durable.

Winnow—to remove chaff from seeds, usually achieved by the pouring of the seed/chaff mixture between buckets with a wind current. The lighter chaff blows away while seeds fall.

References

Alteri, Miguel. 1995. *Agroecology: The science of sustainable agriculture.* 2nd ed. Boulder, CO: Westview Press.

Ashworth, S. 2002. *Seed to seed: Seed saving and growing techniques for vegetable gardeners.* Decorah, IA: Seed Savers Exchange.

Bellon, M. R., J.-L. Pham, and M. T. Jackson. 1997. Genetic conservation: A role for rice farmers. In *Plant genetic conservation,* ed. N. Maxted, B. V. Ford-Lloyd, J. G. Hawkes. London: Chapman and Hall.

Buchmann, S. L., and G. P. Nabhan. 1996. *The forgotten pollinators.* Washington, DC: Island Press/Shearwater Books.

Diamond, J. 1997. *Guns, germs, and steel: The fates of human societies.* New York: W. W. Norton.

Engeland, Ron. 1991. *Growing great garlic.* Okanogan, WA: Filaree Publications.

George, R. A. T. 1999. *Vegetable seed production.* 2nd ed. New York: CABI Publishing.

Goldman, A. 2004. *The compleat squash: A passionate grower's guide to pumpkins, squashes, and gourds.* New York: Artisan.

Heiser, C. B. 1969. *The fascinating world of the nightshades: Tobacco, mandrake, potato, tomato, pepper, eggplant, etc.* New York: Dover Publications, Inc.

———. 1985. *Of plants and people.* Norman, OK: University of Oklahoma Press.

Johnston, R. 1983. *Growing garden seeds.* Albion, ME: Johnny's Select Seeds.

Kloppenburg, J. R. 1988. *First the seed: The political economy of plant biotechnology, 1492–2000.* Cambridge: Cambridge University Press.

Lawn, C. R., and E. Rogosa, ed. 2004. *Restoring our seed* (NE-SARE).

Long, C., and L. Reiley. 2004. Is agrobusiness making our food less nutritious? *Mother Earth News,* June/July.

Meffe, G. K. and C. R. Carrol. 1997. *Principles of conservation biology,* 2nd ed. Sunderland, MA: Sinauer.

Merrick, L. C. 1990. Systematics and evolution of a domesticated squash, *Cucurbita argyrosperma,* and its wild and weedy relatives. In *Biology and utilization of the* Cucurbitaceae, ed. D. M. Bates, R. W. Robinson, C. Jaffrey. Ithaca, NY: Comstock Publishing Associates, Cornell University.

Nabhan, G. P. 1989. *Enduring seeds: Native american agriculture and wild plant conservation.* Berkeley, CA: North Point Press.

———. 1993. *Songbirds, truffles, and wolves: An American naturalist in Italy*. New York: Penguin.

———. 2002. *Coming home to eat: The pleasures and politics of local foods*. New York: W. W. Norton.

Public Seed Initiative (PSI). 2004. *Seed production and seed cleaning workbook*. Ithaca, NY: Cornell University.

Robinson, R. 1996. *Return to resistance: Breeding crops to reduce pesticide dependence*. Davis, CA: AgAccess.

Shiva, V. 2000. *Stolen harvest: The hijacking of the global food supply*. Cambridge, MA: South End Press.

Simplot Company. 1985. *Potato field manual: Nutrient deficiencies, diseases and insect damage symptoms*. Pocatello, ID. (Available from Fedco.)

Simpson, B. B., and M. C. Ogorzaly. 2001. *Economic botany: Plants in our world*. 3rd ed. Boston: McGraw Hill.

Smith, B. D. 1995. *The emergence of agriculture*. New York: Scientific American Library.

Stearns, T. 2004. How to produce cucurbit seed. In C. R. Lawn and E. Rogosa, *Restoring our seed* (NE-SARE).

Whealy, K. 1999. *Garden seed inventory*. 5th ed. Decorah, IA: Seed Savers Exchange.

White, R. M. 2002. "Haudensaunee Native Seed Conservation." Hadley, MA: Hampshire College.

Web Sites:

http://btny.purdue.edu/extension

http://ohioline.osu.edu/b672/index.html

www.ams.usda.gov/science/PVPO/PVPO_Act/PVPA.htm

www.ars-grin.gov/npgs/

www.growseed.org

www.kokopelli-seed-foundation.com

www.ndsu.nodak.edu/soybeandiseases/

www.plbr.cornell.edu/psi

www.statenj.us/agriculture/news/plo309a.htm

Further Reading
(Adapted from the PSI Web site)

Agrawal, R. 1998, *Fundamentals of plant breeding and hybrid seed production.* Enfield, NH: Science Publishers.

Allard, Robert. 1999. *Principles of plant breeding,* 2nd edition. New York: John Wiley & Sons.

Ausubel, Kenny. 1994. *Seeds of change.* New York: HarperCollins.

Bubel, Nancy. 1988. *The new seed starters handbook.* Emmaus, PA: Rodale Press.

Burbank, Luther. 1921. *How plants are trained to work for man: Plant breeding.* New York: P. F. Collier & Son Company.

Burr, Fearing. 1994. *Field and garden vegetables of America.* Chillicothe, IL:.
· The America Botanist Booksellers.

Coleman, Eliot. 1999. *Four-season harvest,* 2nd edition. White River Junction, VT: Chelsea Green Publishing.

Coleman, Eliot. 1995. *The new organic grower,* 2nd edition. White River Junction, VT: Chelsea Green Publishing.

DeBoef, W. 1993. *Cultivating knowledge.* Sterling, VA: Stylus Publishing.

Deppe, Carol. 2000. *Breed your own vegetable varieties.* White River Junction, VT: Chelsea Green Publishing.

Facciola, Stephen. 1990. *Cornucopia: A source book of edible plants.* Vista, CA: Kampong Publications.

Fowler, Cary, and Pat Mooney. 1990. *Shattering: Food, politics, and the loss of genetic diversity.* Tucson, AZ: University of Arizona Press.

Jabs, Carolyn. 1984. *The heirloom gardener.* San Francisco: Sierra Club Books.

Ogden, Shepherd. 1999. *Straight-ahead organic.* White River Junction, VT: Chelsea Green Publishing.

Poehlman, J., and D. Sleper. 1995. *Breeding field crops.* Ames, IA: Iowa State University Press.

Rechcigl, Nancy A., and Jack E. Rechcigl, eds. 1997. *Environmentally safe approaches to crop disease control.* Boca Raton, FL: CRC Press.

Rogers, Marc. 1978. *Growing & storing vegetable seeds.* Charlotte, VT: Garden Way Publishing.

Simmonds, N., and J. Smartt (eds.). 1999. *Principles of crop improvement.* New York: John Wiley & Sons/Blackwell Science.

Smith, Edward C. 2000. *The vegetable gardener's bible. Edward C. Smith.* North Adams, MA: Storey Publishing.

Strickland, Sue, Kent Whealy, and David Cavagnaro (photographer). 1998. *Heirloom vegetables: A home gardener's guide to finding and growing vegetables from the past.* New York: Simon & Schuster.

Turner, Carole B. 1998. *Seed sowing and saving.* North Adams, MA: Storey Publishing.

Index

About the Author and Illustrator

Bryan Connolly farms organically at his small homestead in northeastern Connecticut where he buys and grows seeds from far and wide to satisfy his fascination with plant genetics. When not "swamping the world with plant propagules," Bryan raises goats, chickens, and turkeys on the farm and assists customers at the Willimantic Food Co-op. He has a master's degree in pollination biology from the University of Connecticut and has worked for the University of Mississippi's Medicinal Plant Garden, New England Wild Flower Society, and the Connecticut Department of Environmental Protection.

Contributing editor C. R. Lawn is the founder of Fedco Seeds Cooperative (in 1978), and has worked for the cooperative since. Every year he writes the seeds part of the Fedco catalog. He is co-coordinator, with Elisheva Rogosa, of the SARE-funded Restoring Our Seed Project aimed at improving the quality and quantity of farmer seed production in the Northeast. He lives in Colrain, Massachusetts.

Illustrator Jocelyn Langer is an artist, music teacher, and organic gardener, and the illustrator of the NOFA organic farming handbooks. She illustrates and does graphic design work for alternative media and political events as well as organic-farming-related publications. Jocelyn lives in central Massachusetts.

The photos in the text, with the exception of page 26, are by Bryan Connolly. The special farmer-reviewer for this manual was Michael Glos, and the scientific reviewer was Mark Hutton.

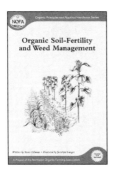

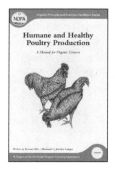
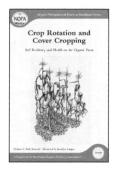